"Until I read Michael Pollan's original, provocative and charming *The Botany of Desire*, I had never managed to get inside the soul of a plant. Mr. Pollan, an accomplished gardener and garden writer, presents a plant's-eye view of the world that challenged some of my most basic assumptions about gardening, particularly the one about whether I control my lilies or they control me. . . . Mr. Pollan disabused me of my anthropocentric ignorance. . . . In his elegant sections on marijuana and potatoes, Mr. Pollan braids together cosmic ideas, conversations with experts and day-to-day reports from his own garden. . . . Mr. Pollan's discussion of the genetically engineered NewLeaf potato, which was devised to resist its most dreaded enemy, the Colorado potato beetle, is a lucid and balanced assessment of this new horticultural technology, a subject too often tackled with barely muffled hysteria."
— *The Wall Street Journal*

"A don't-wanna-put-it-down unspooling of the socio-political, economic and historical forces that led to the cultivation of four crops . . . Pollan is a master at making connections, seeing the lines that connect disparate dots in the complexities of the garden, be they of a political, literary, historical, socioeconomic or, even, sexual realm."
— *Chicago Tribune*

"Funny, interesting and as delicious as a slice of summer peach . . . a must for people who like a good story."
—New York *Daily News*

"An apple, a potato, the tulip and marijuana: This witty book tells the story of each from the plant's point of view."
—*People*

"Engaging, insightful and copiously researched . . . an edifying journey through science and history by way of four plants."
—*Hartford Courant*

"*The Botany of Desire* is an immensely readable, thought-provoking and unusual—indeed uncategorizable—book. . . . Most garden writers could not begin to write such a book."
—*The New York Observer*

"You can trust Pollan to introduce you to a world in your own back yard that you never knew existed, and to make you think about it in a way you never before imagined."
—*St. Louis Post-Dispatch*

"In luminous and often funny prose, Pollan illustrates how these species evolved to satisfy our most basic cravings—and, by doing so, became indispensable."
—*Vogue*

"[Pollan's] insight into Holland's seventeenth-century tulip trade, in which vast sums were bet on those gnarled bulbs, so beautifully analogizes the Internet bubble that it's a perfect object lesson for how it's never 'different this time.' "
—*Esquire*

"Using accessible science, historical events and personal anecdotes, Pollan tells the story of how humans have manipulated apples, tulips, cannabis and potatoes and how the plants have exploited us."
—*Home & Garden*

"A surprising and sometimes shocking look at how our basic human desires have shaped the evolution of four very different plants."
—*Organic Gardening*

MICHAEL POLLAN is the author of seven books, including *Cooked: The Natural History of Transformation, Food Rules, In Defense of Food,* and *The Omnivore's Dilemma.* A longtime contributor to *The New York Times,* he is also the Knight Professor of Science and Environmental Journalism at the University of California, Berkeley. In 2010, *Time* magazine named him one of the one hundred most influential people in the world.

By Michael Pollan

Cooked: A Natural History of Transformation

Food Rules: An Eater's Manual

In Defense of Food: An Eater's Manifesto

The Omnivore's Dilemma: A Natural History of Four Meals

The Botany of Desire: A Plant's-Eye View of the World

A Place of My Own: The Education of an Amateur Builder

Second Nature: A Gardener's Education

THE

BOTANY

OF

DESIRE

RANDOM HOUSE

TRADE PAPERBACKS

NEW YORK

THE

BOTANY

OF

DESIRE

A Plant's-Eye View of the World

MICHAEL

POLLAN

This work was originally published in hardcover by Random House, an imprint of
The Random House Publishing Group, a division of Random House, Inc., in 2001.

Library of Congress Cataloging-in-Publication Data
Pollan, Michael.
The botany of desire : a plant's-eye view of the world / Michael Pollan.
p. cm.
Includes bibliographical references (p.).
ISBN 0-375-76039-3
1. Human-plant relationships. I. Title.
QK46.5.H85 P66   2001   306.4'5—dc21   00-066479

Random House website address: www.atrandom.com
Printed in the United States of America
42 43 44 45 46 47 48 49 50

Book design by J. K. Lambert

For my parents,

who never doubted (or, if they did, never let it show);

and my grandfather, with gratitude

CONTENTS

ACKNOWLEDGMENTS

I had a great deal of help in the making of this book at every step of the way. My thanks first to all the people who gave so generously of their time and knowledge while I was reporting and researching the project; their names appear in the Sources.

Ever since I started writing books a dozen or so years ago, I've had the privilege and even greater pleasure of working with Ann Godoff; indeed, by now I can't imagine writing a book without the net of her wisdom, trust, and friendship. My literary agent, Amanda Urban, has also been there since the beginning. She knew before anyone else that *The Botany of Desire* was the book I should be writing, and, straight through, her judgment on all matters has been indispensable.

Mark Edmundson has also had a hand in all three of my books, though for no other reason but friendship. He read the manuscript with great care and intelligence, parts of it more than once, and every page he touched he made better. Just as important, though, have been the sympathetic ear and priceless reading suggestions he offered along the way. I've been incredibly fortunate, too, to have the gifted editorial eye of Paul Tough, who has gracefully morphed from student to teacher; his suggestions were invaluable. I also owe a large debt of gratitude to Mardi Mellon, at the Union of Concerned Scientists, who generously brought her scientific eye to bear on the manuscript, saving me from all manner of embarrassment; whatever errors remain, however, are mine alone.

My initial forays into the worlds of marijuana-growing and genetically engineered potatoes were sponsored by *The New York Times Magazine;* heartfelt thanks to Gerry Marzorati, Adam Moss, and Jack Rosenthal for their unstinting support and encouragement, as well as to Stephen Mihm for his stellar research assistance. Carol Schneider, Robbin Schiff, Benjamin Dreyer, Alexa Cassanos, and Kate Niedzwiecki have been invaluable allies, as are, always, Jack Hitt, Mark Danner, and Allan Gurganus. Thanks also to Isaac Pollan for his encouragement and, on the bad days, his understanding and comfort.

And finally, to Judith, who really comes first, because without her eye, ear, wisdom, support, patience, encouragement, discernment, foresight, confidence, companionship, judgment, clarity, humor, and love, none of this would ever have gotten done.

Cornwall Bridge, Connecticut
October 2000

INTRODUCTION

The Human Bumblebee

The seeds of this book were first planted in my garden—while I was planting seeds, as a matter of fact. Sowing seed is pleasant, desultory, not terribly challenging work; there's plenty of space left over for thinking about other things while you're doing it. On this particular May afternoon, I happened to be sowing rows in the neighborhood of a flowering apple tree that was fairly vibrating with bees. And what I found myself thinking about was this: What existential difference is there between the human being's role in this (or any) garden and the bumblebee's?

If this sounds like a laughable comparison, consider what it was I was doing in the garden that afternoon: disseminating the genes of one species and not another, in this case a fingerling potato instead of, let's say, a leek. Gardeners like me tend to think such choices are our sovereign prerogative: in the space of this garden, I tell myself, I alone determine which species will thrive and which will disappear. I'm in charge here, in other words, and behind me stand other humans still more in charge: the long chain of gardeners and botanists, plant breeders, and, these days, genetic engineers who "selected," "developed," or "bred" the particular potato that I decided to plant. Even our grammar makes the terms of this relationship perfectly clear: *I choose the plants, I pull the weeds, I harvest the crops.* We divide the world into subjects and objects, and here in the garden, as in nature generally, we humans are the subjects.

But that afternoon in the garden I found myself wondering: What if that grammar is all wrong? What if it's really nothing more than a self-serving conceit? A bumblebee would probably also regard himself as a subject in the garden and the bloom he's plundering for its drop of nectar as an object. But we know that this is just a failure of his imagination. The truth of the matter is that the flower has cleverly manipulated the bee into hauling its pollen from blossom to blossom.

The ancient relationship between bees and flowers is a classic example of what is known as "coevolution." In a coevolutionary bargain like the one struck by the bee and the apple tree, the two parties act on each other to advance their individual interests but wind up trading favors: food for the bee, transportation for the apple genes. Consciousness needn't enter into it on either side, and the traditional distinction between subject and object is meaningless.

Matters between me and the spud I was planting, I realized, really aren't much different; we, too, are partners in a coevolutionary relationship, as indeed we have been ever since the birth of agriculture more than ten thousand years ago. Like the apple blossom, whose form and scent have been selected by bees over countless generations, the size and taste of the potato have been selected over countless generations by us—by Incas and Irishmen, even by people like me ordering french fries at McDonald's. Bees and humans alike have their criteria for selection: symmetry and sweetness in the case of the bee; heft and nutritional value in the case of the potato-eating human. The fact that one of us has evolved to become intermittently aware of its desires makes no difference whatsoever to the flower or the potato taking part in this arrangement. All those plants care about is what every being cares about on the most basic genetic level: making more copies of itself. Through trial and error these plant species have found that the best way to do that is to induce animals—bees or people, it hardly matters—to spread their genes. How? By playing on the animals' desires, conscious and otherwise. The flowers and spuds that manage to do this most effectively are the ones that get to be fruitful and multiply.

So the question arose in my mind that day: Did I choose to plant these potatoes, or did the potato make me do it? In fact, both statements are true. I can remember the exact moment that spud seduced me, showing off its knobby charms in the pages of a seed catalog. I think it was the tasty-sounding "buttery yellow flesh" that did it. This was a trivial, semiconscious event; it never occurred to me that our catalog encounter was of any evolutionary consequence whatsoever. Yet evolution consists of an infinitude of trivial, unconscious events, and in the evolution of the potato my reading of a particular seed catalog on a particular January evening counts as one of them.

That May afternoon, the garden suddenly appeared before me in a whole new light, the manifold delights it offered to the eye and nose and tongue no longer quite so innocent or passive. All these plants, which I'd always regarded as the objects of my desire, were also, I realized, subjects, acting on me, getting me to do things for them they couldn't do for themselves.

And that's when I had the idea: What would happen if we

looked at the world beyond the garden this way, regarded our place in nature from the same upside-down perspective?

This book attempts to do just that, by telling the story of four familiar plants—the apple, the tulip, cannabis, and the potato—and the human desires that link their destinies to our own. Its broader subject is the complex reciprocal relationship between the human and natural world, which I approach from a somewhat unconventional angle: I take seriously the plant's point of view.

⠇⠇

The four plants whose stories this book tells are what we call "domesticated species," a rather one-sided term—that grammar again—that leaves the erroneous impression that we're in charge. We automatically think of domestication as something we do to other species, but it makes just as much sense to think of it as something certain plants and animals have done to us, a clever evolutionary strategy for advancing their own interests. The species that have spent the last ten thousand or so years figuring out how best to feed, heal, clothe, intoxicate, and otherwise delight us have made themselves some of nature's greatest success stories.

The surprising thing is, we don't ordinarily regard species like the cow and the potato, the tulip and the dog, as nature's more extraordinary creatures. Domesticated species don't command our respect the way their wild cousins often do. Evolution may reward interdependence, but our thinking selves continue to prize self-reliance. The wolf is somehow more impressive to us than the dog.

Yet there are fifty million dogs in America today, only ten thousand wolves. So what does the dog know about getting along in this world that its wild ancestor doesn't? The big thing the dog knows about—the subject it has mastered in the ten thousand years it has been evolving at our side—is us: our needs and desires, our emotions and values, all of which it has folded into its genes as part of a sophisticated strategy for survival. If you could read the genome of the dog like a book, you would learn a great deal about who we are and what makes us tick. We don't ordinarily give plants as much credit as animals, but the same would be true of the genetic books of the apple, the tulip, cannabis, and the potato. We could read volumes about ourselves in their pages, in the ingenious sets of instructions they've developed for turning people into bees.

After ten thousand years of coevolution, their genes are rich archives of cultural as well as natural information. The DNA of that tulip there, the ivory one with the petals attenuated like sabers, contains detailed instructions on how best to catch the eye not of a bee but of an Ottoman Turk; it has something to tell us about that age's idea of beauty. Likewise, every Russet Burbank potato holds within it a treatise about our industrial food chain—and our taste for long, perfectly golden french fries. That's because we have spent the last few thousand years remaking these species through artificial selection, transforming a tiny, toxic root node into a fat, nourishing potato and a short, unprepossessing wildflower into a tall, ravishing tulip. What is much less obvious, at least to us, is that these plants have, at the same time, been going about the business of remaking us.

::.

I call this book *The Botany of Desire* because it is as much about the human desires that connect us to these plants as it is about the plants themselves. My premise is that these human desires form a part of natural history in the same way the hummingbird's love of red does, or the ant's taste for the aphid's honeydew. I think of them as the human equivalent of nectar. So while the book explores the social history of these plants, weaving them into our story, it is at the same time a natural history of the four human desires these plants evolved to stir and gratify.

I'm interested not only in how the potato altered the course of European history or how cannabis helped fire the romantic revolution in the West, but also in the way notions in the minds of men and women transformed the appearance, taste, and mental effects of these plants. Through the process of coevolution human ideas find their way into natural facts: the contours of a tulip's petals, say, or the precise tang of a Jonagold apple.

The four desires I explore here are *sweetness*, broadly defined, in the story of the apple; *beauty* in the tulip's; *intoxication* in the story of cannabis; and *control* in the story of the potato—specifically, in the story of a genetically altered potato I grew in my garden to see where the ancient arts of domestication may now be headed. These four plants have something important to teach us about these four desires—that is, about what makes us tick. For instance, I don't think we can begin to understand beauty's gravitational pull without first understanding the flower, since it was the flower that first ushered the idea of beauty into the world the moment, long ago, when floral attraction emerged as an evolutionary strategy. By the same token, intoxication is a human desire we might never have cultivated had it not been for a handful of plants that manage to manufacture chemicals with the precise molecular key needed to unlock the mechanisms in our brain governing pleasure, memory, and maybe even transcendence.

Domestication is about a whole lot more than fat tubers and docile sheep; the offspring of the ancient marriage of plants and people are far stranger and more marvelous than we realize. There is a natural history of the human imagination, of beauty, religion, and possibly philosophy too. One of my aims in this book is to shed some light on the part in that history these ordinary plants have played.

∷

Plants are so unlike people that it's very difficult for us to appreciate fully their complexity and sophistication. Yet plants have been evolving much, much longer than we have, have been inventing new strategies for survival and perfecting their designs for so long that to say that one of us is the more "advanced" really depends on how you define that term, on what "advances" you value. Naturally we value abilities such as consciousness, toolmaking, and language, if only because these have been the destinations of our own evolutionary journey thus far. Plants have traveled all that distance and then some—they've just traveled in a different direction.

Plants are nature's alchemists, expert at transforming water, soil, and sunlight into an array of precious substances, many of them beyond the ability of human beings to conceive, much less manufacture. While we were nailing down consciousness and learning to walk on two feet, they were, by the same process of natural selection, inventing photosynthesis (the astonishing trick of converting sunlight into food) and perfecting organic chemistry. As it turns out, many of the plants' discoveries in chemistry and physics have served us well. From plants come chemical compounds that nourish and heal and poison and delight the senses, others that rouse and put to sleep and intoxicate, and a few with the astounding power to alter consciousness—even to plant dreams in the brains of awake humans.

Why would they go to all this trouble? Why should plants bother to devise the recipes for so many complex molecules and then expend the energy needed to manufacture them? One important reason is defense. A great many of the chemicals plants produce are designed, by natural selection, to compel other creatures to leave them alone: deadly poisons, foul flavors, toxins to confound the minds of predators. But many other of the substances plants make have exactly the opposite effect, drawing other creatures to them by stirring and gratifying their desires.

The same great existential fact of plant life explains why plants make chemicals to both repel and attract other species: immobility. The one big thing plants can't do is move, or, to be more precise, locomote. Plants can't escape the creatures that prey on them; they also can't change location or extend their range without help.

And so about a hundred million years ago plants stumbled on a way—actually a few thousand different ways—of getting animals to carry them, and their genes, here and there. This was the evolutionary watershed associated with the advent of the angiosperms, an extraordinary new class of plants that made showy flowers and formed large seeds that other species were induced to disseminate. Plants began evolving burrs that attach to animal fur like Velcro, flowers that seduce honeybees in order to powder their thighs with pollen, and acorns that squirrels obligingly taxi from one forest to another, bury, and then, just often enough, forget to eat.

Even evolution evolves. About ten thousand years ago the world witnessed a second flowering of plant diversity that we would come to call, somewhat self-centeredly, "the invention of agriculture." A group of angiosperms refined their basic put-the-animals-to-work strategy to take advantage of one particular animal that had evolved not only to move freely around the earth, but to think and trade complicated thoughts. These plants hit on a remarkably clever strategy: getting us to move and think for them. Now came edible grasses (such as wheat and corn) that incited humans to cut down vast forests to make more room for them; flowers whose beauty would transfix whole cultures; plants so compelling and useful and tasty they would inspire human beings to seed, transport, extol, and even write books about them. This is one of those books.

So am I suggesting that the plants made me do it? Only in the sense that the flower "makes" the bee pay it a visit. Evolution doesn't depend on will or intention to work; it is, almost by definition, an unconscious, unwilled process. All it requires are beings compelled, as all plants and animals are, to make more of themselves by whatever means trial and error present. Sometimes an adaptive trait is so clever it appears purposeful: the ant that "cultivates" its own gardens of edible fungus, for instance, or the pitcher plant that "convinces" a fly it's a piece of rotting meat. But such traits are clever only in retrospect. Design in nature is but a concatenation of accidents, culled by natural selection until the result is so beautiful or effective as to seem a miracle of purpose.

By the same token, we're prone to overestimate our own agency in nature. Many of the activities humans like to think they undertake for their own good purposes—inventing agriculture, outlawing certain plants, writing books in praise of others—are mere contingencies as far as nature is concerned. Our desires are simply more grist for evolution's mill, no different from a change in the weather: a peril for some species, an opportunity for others. Our grammar might teach us to divide the world into active subjects and passive objects, but in a coevolutionary relationship every subject is also an object, every object a subject. That's why it makes just as much sense to think of agriculture as something the grasses did to people as a way to conquer the trees.

∷

When Charles Darwin was writing *The Origin of Species,* deciding how best to spring his outlandish idea of natural selection on the world, he settled on a curious rhetorical strategy. Rather than open the book with an account of his new theory, he began with a side subject he judged people (and perhaps English gardeners in particular) would have an easier time getting their heads around. Darwin devoted the first chapter of *The Origin of Species* to a special case of natural selection called "artificial selection"—his term for the process by which domesticated species come into the world. Darwin was using the word *artificial* not as in *fake* but as in *artifact:* a thing reflecting human will. There's nothing fake about a hybrid rose or a butter pear, a cocker spaniel or a show pigeon.

These were a few of the domesticated species Darwin wrote about in his opening chapter, demonstrating how in each case the species proposes a wealth of variation from which humans then select the traits that will be passed down to future generations. In the special realm of domestication, Darwin explained, human desire (sometimes consciously, sometimes not) plays the same role that blind nature does everywhere else, determining what constitutes "fitness" and thereby leading, over time, to the emergence of new forms of life. The evolutionary rules are the same ("modification by descent"), but Darwin understood that they'd be easier to follow in the story of the tea rose than the sea turtle, in the setting of the garden than the Galápagos.

In the years since Darwin published *The Origin of Species,* the crisp conceptual line that divided artificial from natural selection has blurred. Whereas once humankind exerted its will in the relatively small arena of artificial selection (the arena I think of, metaphorically, as a garden) and nature held sway everywhere else, today the force of our presence is felt everywhere. It has become much harder, in the past century, to tell where the garden leaves off and pure nature begins. We are shaping the evolutionary weather in ways Darwin could never have foreseen; indeed, even the weather itself is in some sense an artifact now, its temperatures and storms the reflection of our actions. For a great many species today, "fitness" means the ability to get along in a world in which humankind has become the most powerful evolutionary force. Artificial selection has become a much more important chapter in natural history as it has moved into the space once ruled exclusively by natural selection.

That space, which is the one we often call "the wild," was never quite as innocent of our influence as we like to think; the Mohawks and Delawares had left their marks on the Ohio wilderness long before John Chapman (aka Johnny Appleseed) showed up and began planting apple trees. Yet even the dream of such a space has become hard to sustain in a time of global warming, ozone holes, and technologies that allow us to modify life at the genetic level—one of the wild's last redoubts. Partly by default, partly by design, all of nature is now in the process of being domesticated—of coming, or finding itself, under the (somewhat leaky) roof of civilization. Indeed, even the wild now depends on civilization for its survival.

Nature's success stories from now on are probably going to look a lot more like the apple's than the panda's or white leopard's. If those last two species have a future, it will be because of human desire; strangely enough, their survival now depends on what amounts to a form of artificial selection. This is the world in which we, along with Earth's other creatures, now must make our uncharted way.

This book takes place in that world; consider it a set of dispatches from Darwin's ever-expanding garden of artificial selection. Its main characters are four of that world's success stories. The dogs, cats, and horses of the plant world, these domesticated species are familiar to everyone, so deeply woven into the fabric of our everyday lives that we scarcely think of them as "species" or parts of "nature" at all. But why is that? I suspect it's at least partly the fault of the word. "Domestic" implies that these species have come in or been brought under civilization's roof, which is true enough; yet the house-y metaphor encourages us to think that by doing so they have, like us, somehow *left* nature, as if nature were something that only happens outside.

This is simply another failure of imagination: nature is not only to be found "out there"; it is also "in here," in the apple and the potato, in the garden and the kitchen, even in the brain of a man beholding the beauty of a tulip or inhaling the smoke from a burning cannabis flower. My wager is that when we can find nature in these sorts of places as readily as we now find it in the wild, we'll have traveled a considerable distance toward understanding our place in the world in the fullness of its complexity and ambiguity.

I've chosen the apple, the tulip, cannabis, and the potato for several logical-sounding reasons. One is that they represent four important classes of domesticated plants (a fruit, a flower, a drug plant, and a staple food). Also, having grown these four plants at one time or another in my own garden, I'm on fairly intimate terms with them. But the real reason I chose these plants and not another four is simpler than that: they have great stories to tell.

Each of the chapters that follows takes the form of a journey that either starts out, stops by, or ends up in my garden but along the way ventures far afield, both in space and historical time: to seventeenth-century Amsterdam, where, for a brief, perverse moment, the tulip became more precious than gold; to a corporate campus in St. Louis, where genetic engineers are reinventing the potato; and back to Amsterdam, where another, far less lovely flower has made itself, again, more precious than gold. I also travel to potato farms in Idaho; follow my species' passion for intoxicating plants down through history and into contemporary neuroscience; and paddle a canoe down a river in central Ohio in search of the real Johnny Appleseed. Hoping to render our relationships with these four species in all their complexity, I look at them, by turns, through a variety of lenses: social and natural history, science, journalism, biography, mythology, philosophy, and memoir.

These are stories, then, about Man and Nature. We've been telling ourselves such stories forever, as a way of making sense of what we call our "relationship to nature"—to borrow that curious, revealing phrase. (What other species can even be said to have a "relationship" to nature?) For a long time now, the Man in these stories has gazed at Nature across a gulf of awe or mystery or shame. Even when the tenor of these narratives changes, as it has over time, the gulf remains. There's the old heroic story, where Man is at war with Nature; the romantic version, where Man merges spiritually with Nature (usually with some help from the pathetic fallacy); and, more recently, the environmental morality tale, in which Nature pays Man back for his transgressions, usually in the coin of disaster—three different narratives (at least), yet all of them share a premise we know to be false but can't seem to shake: that we somehow stand outside, or apart from, nature.

This book tells a different kind of story about Man and Nature, one that aims to put us back in the great reciprocal web that is life on Earth. My hope is that by the time you close its covers, things outside (and inside) will look a little different, so that when you see an apple tree across a road or a tulip across a table, it won't appear quite so alien, so Other. Seeing these plants instead as willing partners in an intimate and reciprocal relationship with us means looking at ourselves a little differently, too: as the objects of other species' designs and desires, as one of the newer bees in Darwin's garden—ingenious, sometimes reckless, and remarkably unselfconscious. Think of this book as that bee's mirror.

CHAPTER 1

Desire: Sweetness

Plant: The Apple

(*MALUS DOMESTICA*)

If you happened to find yourself on the banks of the Ohio River on a particular afternoon in the spring of 1806—somewhere just to the north of Wheeling, West Virginia, say—you would probably have noticed a strange makeshift craft drifting lazily down the river. At the time, this particular stretch of the Ohio, wide and brown and bounded on both sides by steep shoulders of land thick with oaks and hickories, fairly boiled with river traffic, as a ramshackle armada of keelboats and barges ferried settlers from the comparative civilization of Pennsylvania to the wilderness of the Northwest Territory.

The peculiar craft you'd have caught sight of that afternoon consisted of a pair of hollowed-out logs that had been lashed together to form a rough catamaran, a sort of canoe plus sidecar. In one of the dugouts lounged the figure of a skinny man of about thirty, who may or may not have been wearing a burlap coffee sack for a shirt and a tin pot for a hat. According to the man in Jefferson County who deemed the scene worth recording, the fellow in the canoe appeared to be snoozing without a care in the world, evidently trusting in the river to take him wherever it was he wanted to go. The other hull, his sidecar, was riding low in the water under the weight of a small mountain of seeds that had been carefully blanketed with moss and mud to keep them from drying out in the sun.

The fellow snoozing in the canoe was John Chapman, already well known to people in Ohio by his nickname: Johnny Appleseed. He was on his way to Marietta, where the Muskingum River pokes a big hole into the Ohio's northern bank, pointing straight into the heart of the Northwest Territory. Chapman's plan was to plant a tree nursery along one of that river's as-yet-unsettled tributaries, which drain the fertile, thickly forested hills of central Ohio as far north as Mansfield. In all likelihood, Chapman was coming from Allegheny County in western Pennsylvania, to which he returned each year to collect apple seeds, separating them out from the fragrant mounds of pomace that rose by the back door of every cider mill. A single bushel of apple seeds would have been enough to plant more than three hundred thousand trees; there's no way of telling how many bushels of seed Chapman had in tow that day, but it's safe to say his catamaran was bearing several whole orchards into the wilderness.

The image of John Chapman and his heap of apple seeds riding together down the Ohio has stayed with me since I first came across it a few years ago in an out-of-print biography. The scene, for me, has the resonance of myth—a myth about how plants and people learned to use each other, each doing for the other things they could not do for themselves, in the bargain changing each other and improving their common lot.

Henry David Thoreau once wrote that "it is remarkable how

closely the history of the apple tree is connected with that of man," and much of the American chapter of that story can be teased out of Chapman's story. It's the story of how pioneers like him helped domesticate the frontier by seeding it with Old World plants. "Exotics," we're apt to call these species today in disparagement, yet without them the American wilderness might never have become a home. What did the apple get in return? A golden age: untold new varieties and half a world of new habitat.

As an emblem of the marriage between people and plants, the design of Chapman's peculiar craft strikes me as just right, implying as it does a relation of parity and reciprocal exchange between its two passengers. More than most of us do, Chapman seems to have had a knack for looking at the world from the plants' point of view—"pomocentrically," you might say. He understood he was working for the apples as much as they were working for him. Perhaps that's why he sometimes likened himself to a bumblebee, and why he would rig up his boat the way he did. Instead of towing his shipment of seeds behind him, Chapman lashed the two hulls together so they would travel down the river side by side.

We give ourselves altogether too much credit in our dealings with other species. Even the power over nature that domestication supposedly represents is overstated. It takes two to perform that particular dance, after all, and plenty of plants and animals have elected to sit it out. Try as they might, people have never been able to domesticate the oak tree, whose highly nutritious acorns remain far too bitter for humans to eat. Evidently the oak has such a satisfactory arrangement with the squirrel—which obligingly forgets where it has buried every fourth acorn or so (admittedly, the estimate is Beatrix Potter's)—that the tree has never needed to enter into any kind of formal arrangement with us.

The apple has been far more eager to do business with humans, and perhaps nowhere more so than in America. Like generations of other immigrants before and after, the apple has made itself at home here. In fact, the apple did such a convincing job of this that most of us wrongly assume the plant is a native. (Even Ralph Waldo Emerson, who knew a thing or two about natural history, called it "the American fruit.") Yet there is a sense—a biological, not just metaphorical sense—in which this is, or has become, true, for the apple transformed itself when it came to America. Bringing boatloads of seed onto the frontier, Johnny Appleseed had a lot to do with that process, but so did the apple itself. No mere passenger or dependent, the apple is the hero of its own story.

∷

On a summery October afternoon almost two hundred years later, I found myself on the bank of the Ohio River a few miles south of Steubenville, Ohio, at the exact spot where John Chapman is thought to have set foot in the Northwest Territory for the first time. I'd come here to look for him, or at least that's what I thought I was doing. I wanted to find out what I could about the "real" Johnny Appleseed, the historical figure behind the Disneyfied folk hero, as well as about the apples in whose story Chapman played such a pivotal role. I figured it would be a modest piece of historical detective work: I'd track down the sites of Chapman's orchards, follow his footsteps (and canoe wake) from western Pennsylvania through central Ohio into Indiana, see if maybe I could find one of the trees he planted. And I did all that, though I'm not sure it got me that much closer to the *real* John Chapman, a man who by now has been composted beneath a deep sift of myth and legend and wishful thinking. I did find another Johnny Appleseed, however, as well as another apple, both of which had been lost.

Actually, the apples and the man have suffered a similar fate in the years since they journeyed down the Ohio together in Chapman's double-hulled canoe. Both then had the tang of strangeness about them, and both have long since been sweetened beyond recognition. Figures of tart wildness, both have been thoroughly domesticated—Chapman transformed into a benign Saint Francis of the American frontier, the apple into a blemish-free plastic-red saccharine orb. "Sweetness without dimension" is how one pomologist memorably described the Red Delicious; the same might be said of the Johnny Appleseed promulgated by Walt Disney and several generations of American children's book writers. In both cases a cheap, fake sweetness has been substituted for the real thing, though it would take me a while to figure out exactly what that was—the strong desire that bound them one to the other, and to the country that took them in.

⁝⁝

Of the man lounging in the two-hulled canoe, Robert Price, his biographer, wrote that he "had the thick bark of queerness on him." Indeed. A man with no fixed address his entire adult life, Chapman preferred to spend his nights out of doors; one winter he set up house in a hollowed-out sycamore stump outside Defiance, Ohio, where he operated a pair of nurseries. A vegetarian living on the frontier, he deemed it a cruelty to ride a horse or chop down a tree; he once punished his own foot for squashing a worm by throwing away its shoe. He liked best the company of Indians and children—and rumors trailed him to the effect that he'd once been engaged to marry a ten-year-old girl, who'd broken his heart. Price feels compelled to assure his readers that Chapman "was not a *complete* crank." The emphasis is mine.

I'd brought a copy of Price's 1954 biography with me to Ohio, and I relied on its maps to retrace Appleseed's annual migration from western Pennsylvania, in search of seeds, to his far-flung properties in Ohio and, eventually, Indiana. It was Price's account that had led me to the spot where Chapman first crossed the river into Ohio, in a faded, microscopic burg to the south of Steubenville called Brilliant.

It had taken me a while to find the landmark mentioned in Price's book, a stream that emptied into the Ohio called George's Run. No one in Brilliant seemed to have heard of it. Eventually I discovered that the stream had long since been rerouted through a culvert. Today George's Run flows, unseen, through a concrete pipe, passes a used-car dealership, crosses beneath a savagely potholed street, and finally reemerges from the earth halfway down a steep, littered embankment behind a convenience store. From there it contributes its meager trickle to the Ohio.

The residents of Brilliant had urged Chapman to stay and plant a nursery, but by his lights the place was already overdeveloped. Ever since he'd come west from Longmeadow, Massachusetts, in 1797, at the age of twenty-three, Chapman had shied away from settled places, for reasons of both temperament and business. To people in Brilliant, Chapman explained that he preferred to get out ahead of the settlers moving west, and this would become the pattern of his life: planting a nursery on a tract of wilderness he judged ripe for settlement and then waiting. By the time the settlers arrived, he'd have apple trees ready to sell them. In time he would find a local boy to look after his trees, move on, and start the process all over again. By the 1830s John Chapman was operating a chain of nurseries that reached all the way from western Pennsylvania through central Ohio and into Indiana. It was in Fort Wayne that Chapman died in 1845—wearing the infamous coffee sack, some say, yet leaving an estate that included some

1,200 acres of prime real estate. The barefoot crank died a wealthy man.

Sketchy though they were, the biographical facts were enough to make anyone question the saintly Golden Books version of Johnny Appleseed (the child bride?!), but it was a single botanical fact about the seeds themselves that made me realize that his story had been lost, and probably on purpose. The fact, simply, is this: apples don't "come true" from seeds—that is, an apple tree grown from a seed will be a wildling bearing little resemblance to its parent. Anyone who wants edible apples plants grafted trees, for the fruit of seedling apples is almost always inedible—"sour enough," Thoreau once wrote, "to set a squirrel's teeth on edge and make a jay scream." Thoreau claimed to like the taste of such apples, but most of his countrymen judged them good for little but hard cider—and hard cider was the fate of most apples grown in America up until Prohibition. Apples were something people drank. The reason people in Brilliant wanted John Chapman to stay and plant a nursery was the same reason he would soon be welcome in every cabin in Ohio: Johnny Appleseed was bringing the gift of alcohol to the frontier.

The identification of the apple with notions of health and wholesomeness turns out to be a modern invention, part of a public relations campaign dreamed up by the apple industry in the early 1900s to reposition a fruit that the Women's Christian Temperance Union had declared war on. Carry Nation's hatchet, it seems, was meant not just for saloon doors but for chopping down the very apple trees John Chapman had planted by the millions. That hatchet—or at least Prohibition—is probably responsible for the bowdlerizing of Chapman's story. Johnny Appleseed was revered on the frontier for a great many admirable qualities: he was a philanthropist, a healer, an evangelist (of a doctrine veer-ing perilously close to pantheism), a peacemaker with the Indians. Yet as I looked out at the sluggish brown Ohio sliding west, trying to picture the man in rags riding alongside his cargo of cider seeds, I wondered if all the cultural energy spent painting Chapman as a Christian saint wasn't really just an attempt to domesticate a far stranger, more pagan hero. Maybe in Ohio I could catch a glimpse of his former wildness. His and the apple's both.

⠇

Slice an apple through at its equator, and you will find five small chambers arrayed in a perfectly symmetrical starburst—a pentagram. Each of the chambers holds a seed (occasionally two) of such a deep lustrous brown they might have been oiled and polished by a woodworker. Two facts about these seeds are worth noting. First, they contain a small quantity of cyanide, probably a defense the apple evolved to discourage animals from biting into them; they're almost indescribably bitter.

The second, more important fact about those seeds concerns their genetic contents, which are likewise full of surprises. Every seed in that apple, not to mention every seed riding down the Ohio alongside John Chapman, contains the genetic instructions for a completely new and different apple tree, one that, if planted, would bear only the most glancing resemblance to its parents. If not for grafting—the ancient technique of cloning trees—every apple in the world would be its own distinct variety, and it would be impossible to keep a good one going beyond the life span of that particular tree. In the case of the apple, the fruit nearly always falls far from the tree.

The botanical term for this variability is "heterozygosity," and while there are many species that share it (our own included), in the apple the tendency is extreme. More than any other single trait, it is the apple's genetic variability—its ineluctable wildness—that accounts for its ability to make itself at home in places as different from one another as New England and New Zealand, Kazakhstan and California. Wherever the apple tree goes, its offspring propose so many different variations on what it means to be an apple—at least five per apple, several thousand per tree—that a couple of these novelties are almost bound to have whatever qualities it takes to prosper in the tree's adopted home.

::|::

Exactly where the apple started out from has long been a matter of contention among people who have studied these things, but it appears that the ancestor of *Malus domestica*—the domesticated apple—is a wild apple that grows in the mountains of Kazakhstan. In some places there, *Malus sieversii,* as it's known to botanists, is the dominant species in the forest, growing to a height of sixty feet and throwing off each fall a cornucopia of odd, applelike fruits ranging in size from marbles to softballs, in color from yellow and green to red and purple. I've tried to imagine what May in such a forest must look—and smell!—like, or October, with the forest floor a nubby carpet of reds and golds and greens.

The silk route traverses some of these forests, and it seems likely that travelers passing through would have picked the biggest and tastiest of these fruits to take with them on their journey west. Along the way seeds were dropped, wildlings sprouted, and *Malus* hybridized freely with related species, such as the European crab apples, eventually producing millions of novel apple types all through Asia and Europe. Most of these would have yielded un-palatable fruit, though even these trees would have been worth growing for cider or forage.

True domestication had to await the invention of grafting by the Chinese. Sometime in the second millennium B.C., the Chinese discovered that a slip of wood cut from a desirable tree could be notched into the trunk of another tree; once this graft "took," the fruit produced on new wood growing out from that juncture would share the characteristics of its more desirable parent. This technique is what eventually allowed the Greeks and Romans to select and propagate the choicest specimens. At this point the apple seems to have settled down for a while. According to Pliny, the Romans cultivated twenty-three different varieties of apples, some of which they took with them to England. The tiny, oblate Lady apple, which still shows up in markets at Christmastime, is thought to be one of these.

As Thoreau suggested in an 1862 essay in praise of wild apples, this most "civilized" of trees followed the westward course of empire, from the ancient world to Europe and then on to America with the early settlers. Much like the Puritans, who regarded their crossing to America as a kind of baptism or rebirth, the apple couldn't cross the Atlantic without changing its identity—a fact that encouraged generations of Americans to hear echoes of their own story in the story of this fruit. The apple in America became a parable.

The earliest immigrants to America had brought grafted Old World apple trees with them, but in general these trees fared poorly in their new home. Harsh winters killed off many of them outright; the fruit of others was nipped in the bud by late-spring frosts unknown in England. But the colonists also planted seeds, often saved from apples eaten during their Atlantic passage, and these seedling trees, called "pippins," eventually prospered (especially after the colonists imported honeybees to improve pollination, which had been spotty at first). Ben Franklin reported that by 1781 the fame of the Newtown Pippin, a homegrown apple discovered in a Flushing, New York, cider orchard, had already spread to Europe.

In effect, the apple, like the settlers themselves, had to forsake its former domestic life and return to the wild before it could be reborn as an American—as Newtown Pippins and Baldwins, Golden Russets and Jonathans. This is what the seeds on John Chapman's boat were doing. (It may also be what Chapman was doing.) By reverting to wild ways—to sexual reproduction, that is, and going to seed—the apple was able to reach down into its vast store of genes, accumulated over the course of its travels through Asia and Europe, and discover the precise combination of traits required to survive in the New World. The apple probably also found some of what it needed by hybridizing with the wild American crabs, which are the only native American apple trees. Thanks to the species' inherent prodigality, coupled with the work of individuals like John Chapman, in a remarkably short period of time the New World had its own apples, adapted to the soil and climate and day length of North America, apples that were as distinct from the old European stock as the Americans themselves.

∷

From Brilliant, I followed the course of the Ohio down toward Marietta. Moving south, the landscape begins to relax, the steep, rocky hillsides that leap up from the river near Wheeling reclining into rich-looking farmland. It was the first week of October, a Sunday, and many of the cornfields had been only partially shaved, presenting a cartoon of work interrupted. In some fields the tall dun corn had been cut away to reveal an old-time oil derrick. The first oil fields in America were found just outside Marietta; a farmer digging his well would notice bubbles of natural gas percolating through the water—the unmistakable scent of hitting

it big. (Before then, discovering a great apple tree in one's cider orchard had been the ticket.) Most of the oil rigs are stilled and
rusted, but now and then I spotted one still pumping lustily away
as if the year were 1925.

In Marietta, I stopped in at the Campus Martius Museum, a
small brick history museum devoted to Ohio's pioneer days, when
Marietta served as the gateway to the Northwest Territory. The
first thing a visitor encounters is a sprawling tabletop diorama
showing what the area looked like in 1788. That was the year a
Revolutionary War hero named Rufus Putnam, who had won a
charter for his Ohio Company from the Continental Congress, arrived here with a small party of men. Their families would follow
a few months later, after the men had constructed the small walled
settlement that formerly stood on this spot.

Eighteenth-century maps on the walls trace an intricately rami-
fying tree of rivers and streams reaching north from the Muskingum's trunk, connecting the dots of a scatter of place-names
that quickly thins to blankness. The maps force you to think of
Ohio in an unaccustomed way, no longer as a middle but as a beginning, an edge. That, of course, was what this place was in 1801,
when Chapman first stood here: America's threshold place, the
cliff of everything unknown and yet to be—unless, of course, you
happened to be a Delaware or Wyandot, for whom the very notion
of wilderness was an error or a lie. But for a white American in
1801, Marietta was the last stop before stepping over the edge.

:¦:

In 1801 one of the things you could buy in Marietta before head-
ing into the interior was apple trees. Soon after his arrival, Rufus
Putnam had himself planted a nursery on the opposite bank of the
Ohio, so that he might sell trees to the pioneers passing through.
What was surprising about this was that the apples Putnam sold

were not grown from seeds: they were grafted trees. In fact, his nursery offered a selection of the well-known eastern varieties— the Roxbury Russets, Newtown Pippins, and Early Chandlers that had already made their names in colonial New England.

What this meant, of course, was that John Chapman's apples were neither the first in Ohio nor by any stretch the best, for his were seedling trees exclusively. Chapman, somewhat perversely, would have nothing to do with grafted trees. "They can improve the apple in that way," he's supposed to have said, "but that is only a device of man, and it is wicked to cut up trees that way. The correct method is to select good seeds and plant them in good ground and God only can improve the apple."

So what, exactly, was unique about Chapman's operation, and why did it succeed? Apart from his almost fanatical devotion to apples planted from seed, his business was distinguished by its portability: his willingness to pack up and move his apple tree operation to keep pace with the ever-shifting frontier. Like a shrewd real estate developer (which is one way to describe him), Chapman had a sixth sense for exactly where the next wave of development was about to break. There he would go and plant his seeds on a tract of waterfront land (sometimes paid for, sometimes not), confident in the expectation that a few years hence a market for his trees would appear at his doorstep. By the time the settlers came, he'd have two- to three-year-old trees ready for sale at six and a half cents apiece. Chapman was evidently the only appleman on the American frontier pursuing such a strategy. It would have large consequences for both the frontier and the apple.

:||:

If a man had the temperament for it and didn't care about starting a family or putting down roots, selling apple trees along the shifting edge of the frontier was not a bad little business. Apples were precious on the frontier, and Chapman could be sure of a strong demand for his seedlings, even if most of them would yield nothing but spitters. He was selling, cheaply, something everybody wanted—something, in fact, everybody in Ohio needed by law. A land grant in the Northwest Territory specifically required a settler to "set out at least fifty apple or pear trees" as a condition of his deed. The purpose of the rule was to dampen real estate speculation by encouraging homesteaders to put down roots. Since a standard apple tree normally took ten years to fruit, an orchard was a mark of lasting settlement.

An orchard is also an idealized or domesticated version of a forest, and the transformation of a shadowy tract of wilderness into a tidy geometry of apple trees offered a visible, even stirring, proof that a pioneer had mastered the primordial forest. Compared to the awesome majesty of the old-growth trees the early settlers encountered, the modesty of an apple tree, the way it obligingly takes on the forms we give it, holding out its fruit and flowers so near to hand, must have been a tremendous comfort on the frontier.

That's one reason planting an orchard became one of the earliest ceremonies of settlement on the American frontier; the other was the apples themselves. It takes a leap of the historical imagination to appreciate just how much the apple meant to people living two hundred years ago. By comparison, the apple in our eye is a fairly inconsequential thing—a popular fruit (second only to the banana) but nothing we can't imagine living without. It is much harder for us to imagine living without the experience of *sweetness*, however, and sweetness, in the widest, oldest sense, is what the apple offered an American in Chapman's time, the desire it helped gratify.

Sugar was a rarity in eighteenth-century America. Even after cane plantations were established in the Caribbean, it remained a luxury good beyond the reach of most Americans. (Later on, cane sugar became so closely identified with the slave trade that many Americans avoided buying it on principle.) Before the English arrived, and for some time after, there were no honeybees in North America, therefore no honey to speak of; for a sweetener, Indians in the north had relied on maple sugar instead. It wasn't until late in the nineteenth century that sugar became plentiful and cheap enough to enter the lives of very many Americans (and most of them lived on the eastern seaboard); before then the sensation of sweetness in the lives of most people came chiefly from the flesh of fruit. And in America that usually meant the apple.

∷

Sweetness is a desire that starts on the tongue with the sense of taste, but it doesn't end there. Or at least it *didn't* end there, back when the experience of sweetness was so special that the word served as a metaphor for a certain kind of perfection. When writers like Jonathan Swift and Matthew Arnold used the expression "sweetness and light" to name their highest ideal (Swift called them "the two noblest of things"; Arnold, the ultimate aim of civilization), they were drawing on a sense of the word *sweetness* going back to classical times, a sense that has largely been lost to us. The best land was said to be sweet; so were the most pleasing sounds, the most persuasive talk, the loveliest views, the most refined people, and the choicest part of any whole, as when Shakespeare calls spring the "sweet o' the year." Lent by the tongue to all the other sense organs, "sweet," in the somewhat archaic definition of the *Oxford English Dictionary*, is that which "affords enjoyment or gratifies desire." Like a shimmering equal sign, the word *sweetness* denoted a reality commensurate with human desire: it stood for fulfillment.

Since then sweetness has lost much of its power and become

slightly . . . well, saccharine. Who now would think of sweetness as a "noble" quality? At some point during the nineteenth century, a hint of insincerity began to trail the word through literature, and in our time it's usually shadowed by either irony or sentimentality. Overuse probably helped to cheapen the word's power on the tongue, but I think the advent of cheap sugar in Europe, and perhaps especially cane sugar produced by slaves, is what did the most to discount sweetness, both as an experience and as a metaphor. (The final insult came with the invention of synthetic sweeteners.) Both the experience and the metaphor seem to me worth recovering, if for no other reason than to appreciate the apple's former power.

Start with the taste. Imagine a moment when the sensation of honey or sugar on the tongue was an astonishment, a kind of intoxication. The closest I've ever come to recovering such a sense of sweetness was secondhand, though it left a powerful impression on me even so. I'm thinking of my son's first experience of sugar: the icing on the cake at his first birthday. I have only the testimony of Isaac's face to go by (that, and his fierceness to repeat the experience), but it was plain that his first encounter with sugar had intoxicated him—was in fact an ecstasy, in the literal sense of that word. That is, he was beside himself with the pleasure of it, no longer here with me in space and time in quite the same way he had been just a moment before. Between bites Isaac gazed up at me in amazement (he was on my lap, and I was delivering the ambrosial forkfuls to his gaping mouth) as if to exclaim, "Your world contains *this*? From this day forward I shall dedicate my life to it." (Which he basically has done.) And I remember thinking, this is no minor desire, and then wondered: Could it be that sweetness is the prototype of *all* desire?

⫶

Anthropologists have found that cultures vary enormously in their liking for bitter, sour, and salty flavors, but a taste for sweetness appears to be universal. This goes for many animals, too, which shouldn't be surprising, since sugar is the form in which nature stores food energy. As with most mammals, our first experience of sweetness comes with our mother's milk. It could be that we acquire a taste for it at the breast, or we may be born with an instinct for sweet things that makes us desire mother's milk.

Either way, sweetness has proved to be a force in evolution. By encasing their seeds in sugary and nutritious flesh, fruiting plants such as the apple hit on an ingenious way of exploiting the mammalian sweet tooth: in exchange for fructose, the animals provide the seeds with transportation, allowing the plant to expand its range. As parties to this grand coevolutionary bargain, animals with the strongest predilection for sweetness and plants offering the biggest, sweetest fruits prospered together and multiplied, evolving into the species we see, and are, today. As a precaution, the plants took certain steps to protect their seeds from the avidity of their partners: they held off on developing sweetness and color until the seeds had matured completely (before then fruits tend to be inconspicuously green and unpalatable), and in some cases (like the apple's), the plants developed poisons in their seeds to ensure that *only* the sweet flesh is consumed.

Desire, then, is built into the very nature and purpose of fruit, and so, quite often, is a kind of taboo. The vegetable kingdom's lack of glamour by comparison (whoever heard of a forbidden vegetable?) can be laid to the fact that a vegetable's reproductive strategy doesn't turn on turning animals on.

:|:

The blandishments of sugar are what got the apple out of the Kazakh forests, across Europe, to the shores of North America, and eventually into John Chapman's canoe. But the appeal of apples to humans (and perhaps especially to American humans) probably owes to their figurative as well as literal sweetness. The earliest settlers lighting out from places like Marietta wanted apple trees nearby because they were one of the comforts of home. Since the time of the New England Puritans, apples have symbolized, and contributed to, a settled and productive landscape. In the eyes of a European, fruit trees were part and parcel of a sweet landscape, along with clean water, tillable land, and black soil. To call land "sweet" was a way of saying it answered our desires.

The fact that the apple was generally believed to be the fateful tree in the Garden of Eden might also have commended it to a religious people who believed America promised a second Eden. In fact, the Bible never names "the fruit of the tree which is in the midst of the garden," and that part of the world is generally too hot for apples, but at least since the Middle Ages northern Europeans have assumed that the forbidden fruit was an apple. (Some scholars think it was a pomegranate.) This mistake strikes me as yet another example of the apple's gift for insinuating itself into every sort of human environment, even, apparently, a biblical one. Like a botanical Zelig, the apple has wormed its way into our image of Eden through the brushwork of Dürer and Cranach and countless others. After their pictures, re-creating a promised land anywhere in the New World without an apple tree would have been unthinkable.

Especially to a Protestant. There was an old tradition in northern Europe linking the grape, which flourished all through Latin Christendom, with the corruptions of the Catholic Church, while casting the apple as the wholesome fruit of Protestantism. Wine figured in the Eucharist; also, the Old Testament warned against the temptations of the grape. But the Bible didn't have a bad word to say about the apple or even the strong drink that could be made from it. Even the most God-fearing Puritan could persuade himself that cider had been given a theological free pass.

"The desire of the Puritan, distant from help and struggling for bare existence, to add the Pippin to his slender list of comforts, and the sour 'syder' to cheer his heart and liver, must be considered a fortunate circumstance," a speaker told a meeting of the Massachusetts Horticultural Society in 1885. "Perhaps he inclined to cider . . . because it was nowhere spoken against in the scriptures." Whether this was really the reason or a rationale concocted after the fact, Americans were indeed strongly inclined to cider, an inclination that accounts for the high esteem in which the apple was held in the colonies and on the frontier. In fact, there was hardly anything else to drink.

:||:

Alcohol is, of course, the other great beneficence of sugar: it is made by encouraging certain yeasts to dine on the sugars manufactured in plants. (Fermentation converts the glucose in plants into ethyl alcohol and carbon dioxide.) The sweetest fruit makes the strongest drink, and in the north, where grapes didn't do well, that was usually the apple. Up until Prohibition, an apple grown in America was far less likely to be eaten than to wind up in a barrel of cider. ("Hard" cider is a twentieth-century term, redundant before then since virtually all cider was hard until modern refrigeration allowed people to keep sweet cider sweet.)

Corn liquor, or "white lightning," preceded cider on the frontier by a few years, but after the apple trees began to bear fruit, cider—being safer, tastier, and much easier to make—became the alcoholic drink of choice. Just about the only reason to plant an orchard of the sort of seedling apples John Chapman had for sale would have been its intoxicating harvest of drink, available to any-one with a press and a barrel. Allowed to ferment for a few weeks, pressed apple juice yields a mildly alcoholic beverage with about half the strength of wine. For something stronger, the cider can then be distilled into brandy or simply frozen; the intensely alco-holic liquid that refuses to ice is called applejack. Hard cider frozen to thirty degrees below zero yields an applejack of 66 proof.

Virtually every homestead in America had an orchard from which literally thousands of gallons of cider were made every year. In rural areas cider took the place not only of wine and beer but of coffee and tea, juice, and even water. Indeed, in many places cider was consumed more freely than water, even by children, since it was arguably the healthier—because more sanitary—beverage. Cider became so indispensable to rural life that even those who railed against the evil of alcohol made an exception for cider, and the early prohibitionists succeeded mainly in switching drinkers over from grain to apple spirits. Eventually they would attack cider directly and launch their campaign to chop down apple trees, but up until the end of the nineteenth century cider contin-ued to enjoy the theological exemption the Puritans had contrived for it.

It wasn't until this century that the apple acquired its reputa-tion for wholesomeness—"An apple a day keeps the doctor away" was a marketing slogan dreamed up by growers concerned that temperance would cut into sales. In 1900 the horticulturist Lib-erty Hyde Bailey wrote that "the eating of the apple (rather than the drinking of it) has come to be paramount," but for the two centuries before that, whenever an American extolled the virtues of the apple, whether it was John Winthrop or Thomas Jefferson, Henry Ward Beecher or John Chapman, their contemporaries would probably have smiled knowingly, hearing in their words a distinct Dionysian echo that we are apt to miss. When Emerson, for instance, wrote that "man would be more solitary, less friended, less supported, if the land yielded only the useful maize and potato, [and] withheld this ornamental and social fruit," his readers understood it was the support and sociability of alcohol he had in mind.

Americans' "inclination toward cider" is the only way to explain John Chapman's success—how the man could have made a living selling spitters to Ohio settlers when there were already grafted trees bearing edible fruit for sale in Marietta.

::|::

Mount Vernon, Ohio, is a classic early-nineteenth-century town, a modest grid of streets laid out around a central square of green a short walk from the meeting place of two streams. In the library on the square is a map of the town made in 1805, the year it was platted. If you look down in the bottom left-hand corner, where Owl Creek curves in to disturb the tidy grid, you can see lots 145 and 147, both of which were bought by John Chapman in 1809 for the sum of fifty dollars. Follow the creek to the far right-hand edge of the map, and you'll see a neat rank of apple trees, representing what is thought to be one of Chapman's nurseries.

I'd come to Mount Vernon, following the Muskingum and its tributaries north from Marietta, to meet Ohio's leading authority on Johnny Appleseed. William Ellery Jones is a fifty-one-year-old fund-raising consultant and amateur historian with a dream: to establish a Johnny Appleseed Heritage Center and Outdoor Theater on a hillside outside Mansfield. When I had phoned him the month before at his home in Cincinnati, he generously offered to give me a guided tour of "Johnny Appleseed country." Jones hinted that he had made some important discoveries—the locations of various Chapman sites and relics—and he indicated that, if I played my cards right, I might get to see a few of them. This seemed a little too good to be true—finding a Virgil in Appleseed country with one phone call. Three days spent driving around Ohio in the company of this gentle monomaniac confirmed that assessment.

The Heritage Center and Outdoor Theater should have been my tip-off. Within moments of our handshake, I could see that Bill Jones was deeply invested in precisely the version of Chapman's life I'd come west to escape: Saint Appleseed. "Chapman is a hero for our time," he told me in dead earnest when I asked what had attracted him to Chapman's story. "His philanthropy, his self-lessness, his Christian faith. John Chapman was also America's first environmentalist. I ask you, could you *invent* a better role model for our children?" I decided to wait a bit before bringing up the child bride or the applejack.

Jones is a tall, courtly man with pale blue eyes and fine, parchmentlike skin. He gives the impression of being a tightly stretched drum of a man, devoid of irony and, by his own lights, somewhat out of place in time. He's dismayed by present-day America—the popular culture, the violence, the "lack of moral compass." Ohio's frontier past is vividly present to him, and old-timey expressions like "Cripes!," "Gee whillikers!," and "Darn tootin' " come often and unself-consciously to his lips.

One of the first things I noticed about Bill were his delicate white hands and the multiple pairs of leather gloves he carried in his briefcase. Though it was only October, Bill would don the gloves while pumping gas or even indoors, when handed a hot mug of coffee. After we got to know each other a bit, he mentioned that he was certain Chapman had been an obsessive-compulsive, and that people used to make fun of Chapman's delicate hands. "If you didn't have fingers like thumbs back then, folks were liable to say you were effeminate."

Jones had put together an ambitious itinerary between Mount Vernon and Fort Wayne, beginning with a brisk morning walk to plots 145 and 147. Chapman's two landholdings in Mount Vernon stand across the street from each other, on the banks of Owl Creek. Jones said he was applying to the state to have historical markers erected on the site, now the parking lot of a tire dealership. Owl Creek looked far too shallow and sluggish to serve as the busy thoroughfare Jones described, but he pointed out that reservoirs and dams had long ago tamed most of the local streams and rivers. Chapman's Mount Vernon property was, I would discover, typical of his holdings: the land hugged a stream, ensuring water for his seedlings early on and sales traffic later, and they were located on the edge of a new settlement. That particular tropism, pulling Chapman from the center of things out to the margins, proved to be a constant with the man and his life's project.

Over the next few days Bill showed me all over Appleseed country and did an impressive job of bringing that quasi-historical place to life. We traipsed through a dozen of Chapman's former nurseries, pulled over at the drop of a historical marker (Jones deplored the recent switch from brass to aluminum), and stood on a handful of undistinguished street corners that Bill alone knew were "crucial Appleseed sites." On the banks of the Auglaize River we found the site of the famous sycamore stump Chapman had once lived in (now the front lawn of a ranch house), and in a run-down section of Mansfield, we visited the site of his kid sister Persis Broom's house, now a drive-through liquor store called the Galloping Goose. In Defiance we climbed to the top of a water-treatment plant to obtain an unobstructed view of one Appleseed nursery, and near Loudonville we paddled a canoe for two hours to catch a glimpse of another. On a farm outside Savannah, we took pictures of each other standing next to an ancient, half-dead apple tree that may or may not have been planted by Chapman.

All the while Jones spooned tales of Johnny Appleseed, a rich soup of legend sprinkled with chunks of historical and biographical fact. Most of what's known about Chapman comes from accounts left by the many settlers who welcomed him into their cabins, offering the famous appleman/evangelist a meal and a place to sleep. In exchange, his hosts were happy to have Chapman's news (of Indians and Heaven, of his own fantastic exploits) and apple trees (he'd usually plant a couple as a token of his thanks). There was, too, the sheer entertainment value of a guest who was, literally, a legend in his own time.

:|:

Chapman lived everywhere and nowhere. He was constantly on the move, traveling in autumn to Allegheny County to gather seeds, scouting nursery sites and planting in the spring, repairing fences at old nurseries in summer, and, wherever he planted, signing up local agents to keep an eye on and sell his trees, since he was seldom in one place long enough to do that work himself. Even into his sixties, after moving his base of operations to Indiana, Chapman made an annual pilgrimage to central Ohio to look after his nurseries there. His absentee management meant he was frequently gypped, and his land claims were often jumped, though whenever this happened Chapman's main concern seems to have been for the welfare of his trees. In spite of these setbacks, he managed to accumulate enough cash to build up his real estate holdings and give money away to people in need, frequently strangers. As Bill pointed out, the size of his estate—which included some twenty-two parcels of land—is hard to square with the notion that he was feebleminded or feckless.

Even so, he was undoubtedly "one of the oddest characters in our history," as a nineteenth-century historian in Mount Vernon put it. From the reminiscences of the settlers he visited along the route of his annual migration have sprung tall tales of his endurance, generosity, gentleness, heroism, and, it must be said, his unreconstructed strangeness. Jones knows all these tales by heart, and though he is agnostic on the veracity of the tallest ones, he was happy to pass them on—most of them, anyway.

Not surprisingly, Bill dwelled on stories of Chapman's heroism, and together we retraced a portion of the famous "barefoot run" of 1812. During the war with England, Indians allied with the British occasionally rampaged, and late one September night Chapman sprinted thirty miles through the forest from Mansfield to Mount Vernon to warn the settlers of their approach. "Behold, the tribes of the heathens are round about your doors," he's supposed to have cried, "and a devouring flame followeth after them."

As the high-flown diction suggests, Chapman figured himself the hero of a latter-day biblical narrative, a man anointed "to blow the trumpet in the wilderness." This he would blow in every cabin he visited, asking his hosts after supper if they would hear "some news right fresh from Heaven" before producing the Swedenborgian tracts he kept tucked into his waistband. Black eyes blazing, he'd then launch into a sermon fired with a mystic's zeal. Chapman saw himself as a bumblebee on the frontier, bringer of both the seeds and the word of God—of both sweetness, that is, and light.

Swedenborgian doctrine, which holds that everything here on Earth corresponds directly to something in the afterlife, might explain the strange and wonderful ways Chapman conducted himself in nature. The same landscape his countrymen treated as hostile, heathen, and therefore theirs to conquer, Chapman regarded as beneficent in every particular; in his eyes even the lowliest worm glowed with divine purpose. His kindness to animals was notorious, an outrage to frontier custom. It was said he'd sooner douse his campfire than singe the mosquitoes attracted to its flame. Chapman often used his profits to purchase lame horses to save them from slaughter, and once he freed a wolf he found snared in a trap, nursing the animal to health and then keeping it as a pet. When he discovered one evening that the hollowed-out log in which he intended to spend the night was already occupied by bear cubs, he let them be, making his bed in the snow instead. Chapman could sleep anywhere, it seems, though he was partial to hollowed-out logs or a hammock slung between two trees. One time he floated a hundred miles down the Allegheny on a block of ice, sleeping the whole way.

Curiously, a great many stories about Chapman have to do with his feet: how he'd go barefoot in any weather, the time he punished his foot for stepping on a worm (or in some versions a snake). He would entertain boys by pressing needles or hot coals into the soles of his feet, which had grown as horny and tough as an elephant's. (Ripe for ridicule though he must have been, boys were so awed by his fortitude that they never made fun of him.) Listening to an itinerant preacher in Mansfield pound his tree-stump pulpit and ask, one too many times, "Where now is there a man who, like the primitive Christians, is traveling to Heaven barefoot and clad in coarse raiments?" Chapman roused himself from the log he was reclining against and planted his bare ugly foot square in the middle of the preacher's stump. "Here's your primitive Christian!" The recurring barefoot theme underscores the sense people had that Chapman's relationship to nature was special—and not quite human. The soles of our shoes interpose a protective barrier between us and the earth that Chapman had no use for; if shoes are part and parcel of a civilized life, Chapman had one foot planted in another realm, had at least that much in common with the animals. Whenever I hear or read about Chapman's horny bare feet, I can't help picturing him as some kind of satyr or centaur.

But in spite of his peculiar attire and personal habits, those who knew him said he was "never repulsive." People were happy to have him as a guest in their homes, and parents would let him bounce their babies on his lap.

Shadowy stories about Chapman's love life seem to have followed him across the frontier, but whenever I asked Bill Jones about this, he tightened up. In one account Chapman had come west after a girl had stood him up at the altar back in Massachusetts. To people who asked him why he had never married, he would say that God had promised him a "true wife in Heaven" if he would abstain from marriage on Earth. In the most curious of these stories, recounted by Price, Chapman made an arrangement with a frontier family to raise their ten-year-old daughter to be his bride. For several years Chapman paid regular visits to the girl and contributed to her upkeep, until one visit when he chanced to witness his young fiancée flirting with some boys her own age. Wounded and enraged, Chapman abruptly broke off the relationship. True or not, the stories hint at a sexual eccentricity of some kind. Or maybe his libido was submerged in some sort of polymorphous love of nature, as some biographers have theorized about Thoreau.

Gingerly, I tried at one point to raise the subject with Bill. My timing was perhaps not the best. We were in my rental car, driving up the wooded hillside near Mansfield where he hoped someday to build his Heritage Center and "Class A" Outdoor Theater—a destination for school groups and families on vacation, as he'd told me more than once. And here I was, asking whether he thought his hero might have had a . . . thing for young girls.

"I know exactly what story you're referencing," Jones said tightly. "The child bride. In my opinion it is completely implausible."

Jones was silent for a time and then worked himself into a denunciation of "the sort of people who feel compelled to take our heroes down a peg." And then, pulling tensely on the corners of his mouth, he confided his deepest, darkest fear about Chapman, a charge about his hero's sexuality that, though baseless—though never in fact even *alleged* by anyone—nevertheless stood "to ruin everything we're trying to do." I'm sorry to say that the price of hearing this rumor was a promise not to tell.

Jones had his own G-rated theory of Chapman's love life, something having to do with a Massachusetts girl who may have broken a promise to join him in Ohio. "That's as much as I can tell you right now," Bill said. He sounded as if he were talking to Bob Woodward in a parking garage. I pressed, gently. "Nope. Not a word until I can get this all nailed down and published."

That night I went to hear Bill give a talk about Chapman at the Loudonville Historical Society, a stop on his one-man campaign to build support for his Heritage Center and Outdoor Theater. Fifty or so mostly retired folks in folding chairs sipped coffee and listened politely as Jones pressed his case: John Chapman is just the "exemplary figure" to help our children navigate a treacherous world, "yet no one is telling his story." As he spoke, the slide projector showed an early engraving of Chapman made by a woman who had known him in Ohio. Scraggly and barefoot, he's wearing a sackcloth cinched at the waist like a dress and a tin pot on his head; in one hand he's holding out an apple sapling like a scepter. The man looks completely insane.

Bill's talk took the rhetorical form of a sermon, with the line

"No one is telling his story!" serving as its thumping refrain. He was determined to cut Chapman's life to a Christian pattern, and the stories he told were the ones that made the case for beatification. Protoenvironmentalist. Philanthropist. Friend to children and animals and Indians. It was pap, little more, and I wasn't the only person in the room to grow impatient, especially when Jones got around to the apples, which he praised, incredibly, as "an important source of vitamin C on the frontier." Just then an old guy behind me poked an elbow in his neighbor's ribs and whispered, "So does he ever get around to the applejack?"

He did not. Bill was doing lives of the frontier saints, and there was no place in it for alcohol (or mysticism or romance or psychological weirdness of any kind). The sole mention of cider was cider vinegar, "vital as a preservative." (So *that's* who John Chapman was, patron saint of pickling!) Afterward, as we were packing up Bill's tripods and slides, I asked him about the omission. He smiled. "Come on, this is a family show."

⁞

I believe I got a better glimpse of John Chapman the following morning, when Bill and I set out to canoe a stretch of the Mohican north of Loudonville. Bill wanted to show me a riverside nursery of Chapman's, and I was curious to see the country from a perspective nearer to Chapman's own—from the water, I mean, since it was by canoe or pirogue that he usually traveled. On the old maps Chapman carried, the rivers and streams appear as strong black lines against a whole lot of blank space. His America ordered itself around those veiny lines the way ours does around highways. On them you could travel from the spot where Bill and I were starting out all the way to Pittsburgh or to the Mississippi, depending on which way you turned at Marietta.

The sun was not yet up over the trees when we put into the river a few miles above Perrysville, me taking the seat up front since Bill was the more experienced canoer. The water, moving with surprising dispatch for the time of year, looked like a freshly blacktopped road, except where snags flustered its surface, causing it to sparkle. In places spooky mists rose from the surface, and the banks were so thickly lined with trees—giant cottonwoods leaning way out over the water, spectacularly contorted sycamores—that it wasn't hard to pretend we were pushing through a wilderness. In fact, acres of newly cut corn lay just beyond the line of trees, and at one point I glimpsed a chugging factory through an opening in the leaves. We slipped by mergansers and mallards and saw a pileated woodpecker pile-driving the trunk of a dead tree on the bank. At one point a young wood duck allowed us to follow it for at least a hundred feet, probably trying to lead us away from a nest; judging the coast clear, the bird exploded noisily into flight.

After we'd been paddling along for an hour or so, Bill pointed to a broad table of open land off to our left. This was the site of Greentown, a sprawling Indian village Chapman often visited, at least before it was torched by settlers during the War of 1812. Just a few hundred yards farther on, at the spot where a tiny creek dribbled into the river, was the site of Chapman's apple tree nursery. I lifted my paddle, and through the trees I could see a rough stubble of corn on a gently curving skin of black earth.

The nursery's proximity to an Indian town might have troubled another man, but Chapman moved easily between the societies of settler and Native American, even when the two were at war. The Indians regarded Chapman as a brilliant woodsman and medicine man. In addition to the apples, which the Indians were eager for, Chapman brought with him the seeds of a dozen different medicinal plants, including mullein, motherwort, dandelion, wintergreen, pennyroyal, and mayweed, and he was expert in their use.

Chapman's ability to freely cross borders that other people believed to be fixed and unbreachable—between the red world and the white, between wilderness and civilization, even between this world and the next—was one of the hallmarks of his character and probably the thing that most confounded people about the man, both then and now. It certainly confounded me. From a conventional distance, at least, Chapman's whole life appears to have been a skein of warring terms and contradictions no ordinary mind could hope to sustain, much less resolve.

As Bill and I glided slowly down the Mohican, each of us alone with his thoughts, I tried to list some of these contradictions, hoping to discover some pattern. Chapman combined the flinty toughness of a Daniel Boone with the gentleness of a Hindu. He was a deeply pious man—sometimes insufferably so, I imagine ("Will you have some news right fresh from Heaven?")—yet people said he could also enjoy a drink (a pinch of snuff, too) and tell a good joke, often at his own expense. I wondered how he squared the two vocations that occupied his days—that is, bringing to people leading stringent frontier lives two very different kinds of consolation: God's word and hard drink.

The paradoxes piled up. An agent of civilization, working to domesticate the wilderness with his apple trees and herbs and religion, he was at the same time completely at home in the undomesticated wild, as well as in the company of Native Americans, to whom that civilization was toxic. A barefoot backwoodsman draped in sackcloth, Chapman could hold forth knowledgeably on Swedenborgian theology, perhaps the most intellectually demanding religious doctrine of the time.

Maybe that was the key. Maybe it was Swedenborg's thought

that gave Chapman's mind just what it needed to dissolve all these paradoxes. In Swedenborg's philosophy there is no rift between the natural world and the divine. Much like Emerson, who cited him as an influence, Swedenborg claimed that there were one-to-one "correspondences" between natural and spiritual facts, so that close attention and devotion to the former would advance one's understanding of the latter. Thus an apple tree in bloom was part of the natural process of making fruit at the same time it was a "living sermon from God"; likewise, a crow wheeling overhead was a type of the black forces waiting to overtake men's souls when they wandered off the Path. The river before you might be that Path, yet a wrong turn on it might land you in Newark, Ohio, a hard-drinking town notorious for gambling and prostitution that Chapman believed offered a literal preview of Hell. Everything before us was doubled; not this world or that, but both.

Fervently held, such beliefs must have lit up the whole landscape—the rivers and trees, the bears and wolves and crows, even the mosquitoes—with a divine glow. Every Path through the woods was capitalized, every deprivation a spiritual test. Minus the Christian symbolism, I imagine Chapman's was a world much like that inhabited by the ancient Greeks, in which all nature and experience were suffused with divine significance: the storms, the dawns, the strangers at your door. One looked outward, to the land, for meaning, rather than inward or upward.

This was not how nature ordinarily appeared to Americans in Chapman's time. To most of them, the forest was still a heathen chaos. Remember that by the time the New England transcendentalists began to find the divine in nature ("God's second book," they called it), their landscape had been securely under human control for more than a century; Walden Woods was far from a wilderness. For Chapman the natural world even at its wildest was never a falling away or a distraction from the spirit world; it was continuous with it. In some ways this doctrine chimes with the Native Americans' cosmology, which could account for the kinship Chapman felt for Indians and they for him. Chapman's mystical teachings veer about as close to pantheism and nature worship as Christianity has ventured. In Puritan New England he'd have been jailed as a heretic.

It may have been Chapman's conviction that this world is a type or rough draft of the next that allowed him to overlook or dissolve the tensions the rest of us perceive between the realms of matter and spirit, as well as nature and civilization. For him these borders may simply not have been real. So many of the legends about Appleseed depict him as a kind of liminal figure, part man and part . . . well, *something* else. The something else, which was perhaps symbolized by the soles of his bare feet callused to a tough hide, is what permitted him to live with one of those feet planted in our world, the other in the wild. He was a kind of satyr without the sex—a Protestant satyr, you might say, moving through these woods as if they were his true home, making his bed in hollowed logs and his breakfast from a butternut tree, keeping the company of wolves.

As I thought about the scattering of settlers along these streams who would welcome Chapman into their homes, offering a meal and a bed to this strange man in rags, I was reminded of how the gods of classical mythology would sometimes appear at people's doors dressed as beggars. Just to be on the safe side, the Greeks would shower hospitality on even the most dubious stranger, because you never knew when the ragged fellow on your doorstep might turn out to be Athena in disguise. It's true that Johnny Appleseed's fame usually preceded him, but you couldn't blame a settler family for wondering if the man who'd appeared at their door didn't have something otherworldly about him. There was the gleam in his eyes that everyone remarked on, and the news he brought of other worlds (the wild, the Indian, the heavenly); and, of course, there was the precious gift of those apples.

As we glided through the woods to the music of birds and the splish-swirl of our paddles stitching the black water, I tried to summon an image of Chapman. I fell back on one of the slides Bill had shown the night before at the historical society. This one was an etching that had accompanied an 1871 article about Chapman in *Harper's New Monthly Magazine*, and it depicted Chapman as a sinewy, barefoot figure with a goatish beard, wearing, again, something that looks very much like a toga or a dress. The effect was of a creature part man and part woman. Yet it was even more ambiguous than that, since the slight figure with the goat's beard also seems to be melting into, or out of, the shadowy trees all around. What a strange image, I remember thinking, and now I thought I understood why: Chapman appeared in it as a faintly Christianized version of some pagan wood god. And that seemed just about right.

By the time I entertained this little epiphany, the sun had risen high enough behind the trees to flare wildly between the cottonwood leaves, almost like a strobe, momentarily turning the riverscape into a silhouette of itself. I saw Chapman now as clearly as I could hope to. Johnny Appleseed was no Christian saint—that left out too much of who he was, what he stood for in our mythology. Who he was, I realized, was the American Dionysus.

∴

After the river trip my interest in Bill Jones's John Chapman began to flicker. Which was too bad for me, because we still had a lot of ground to cover between here and Fort Wayne, where I planned to catch a flight home. I found myself tuning out a touching story about Chapman buying a new set of china for a family who'd lost all their possessions in a fire. It felt as though there were now two John Chapmans riding with us in the car, Bill's Christian saint and my pagan god, and the front seat began to feel too small for the both of them. This made for an extremely long ride to Fort Wayne.

When at last I got home, I went looking for Appleseed again, this time in the library. I read everything I could find about Dionysus, about whom I knew only the usual high school basics. Teaching men how to ferment the juice of the grape, Dionysus had brought civilization the gift of wine. This was more or less the same gift Johnny Appleseed was bringing to the frontier: because American grapes weren't sweet enough to be fermented successfully, the apple served as the American grape, cider the American wine. But as I delved deeper into the myth of Dionysus, I realized there was much more to his story, and the strangely changeable god who began to come into focus bore a remarkable resemblance to John Chapman. Or at least to "Johnny Appleseed," who, I became convinced, is Dionysus's American son.

Like Johnny Appleseed, Dionysus was a figure of the fluid margins, slipping back and forth between the realms of wildness and civilization, man and woman, man and god, beast and man. I found Dionysus depicted variously as a wild man with foliage sprouting from his head, a goat, a bull, a tree, and a woman. Friedrich Nietzsche paints Dionysus as a figure able to dissolve "all the rigid and hostile barriers" between nature and culture.

The Greeks regarded Dionysus as the antithesis of Apollo, god of clear boundaries, order, and light, of man's firm control over nature. Dionysian revelry melts every Apollonian line, so that, as Nietzsche writes, "alienated, hostile, or subjugated nature . . . celebrates her reconciliation with her lost son, man." By worshiping

Dionysus and getting drunk on his wine, the Athenians temporarily returned to nature, to a time when, as the classicist Jane Harrison writes, "man is still to his own thinking brother of plants and animals." The odd, ecstatic worship of Dionysus, which needed no temple, always took place outside the city, returning religion to the woods where it had begun.

I learned also that Dionysus was the god originally responsible for the marriage of people and plants that John Chapman's double-hulled canoe had symbolized for me. In *The Golden Bough*, James Frazer says that, in addition to the grapevine, Dionysus was also the patron of cultivated trees and specifically credits him with the discovery of the apple. He was in fact the god of domestication itself, bringing "wisdom from the very breast of nature" (Nietzsche), teaching men not only to ferment wine but also to hitch their plow to the ox. Dionysus brought wild plants into the house of civilization, but by the same token his own untamed presence reminded people of the untamed nature on which that house always rests, somewhat unsteadily. The same, I realized, was true of Johnny Appleseed.

Nothing better captures the paradox of Dionysus's double role, as a force for domestication *and* wildness, than his involvement with grapes and wine. Wine itself is a peculiarly liminal substance, poised on the edge of nature and culture as well as civility and abandon. It is truly an extraordinary thing, this artful transformation of raw nature—a fruit!—into a substance with the power to alter human perception. Yet wine is an achievement of civilization we're apt to take for granted or condemn, perhaps especially as Americans, for whom alcohol has always presented a moral problem.

The Greeks, who were much better at holding contradictory ideas in their minds than we are, understood that intoxication could be divine or wretched, a ceremony of human communion or madness, depending on the care taken in handling its magic. "Wine is rudderless," Plato warned. (He advised mixing it with water and serving it in tiny cups.) Dionysian revelry, which begins in ecstasy and often ends in blood, embodies this truth: the same wine that loosens the knots of inhibition and reveals nature's most beneficent face can also dissolve the bonds of civilization and unleash ungovernable passions.

This is why, of all the gods, "Dionysus is, for humans, fiercest and most sweet," according to Euripides. If Apollo is a god of concentrated light, Dionysus, worshiped at night, is a god of dispersed sweetness. Under his influence "the earth flows, flows beneath us, then milk flows, and wine flows, and nectar flows, like flame." Under the spell of Dionysus and his wine, all nature answers to our desires.

As for the fierce part of the Dionysian drama, this Johnny Appleseed did not play. He was a far more gentle, less sexual being than Dionysus, though his gender does sometimes seem equally amorphous. (Come to think of it, Chapman *did* sponsor sex orgies—but only among apple trees.) The flight from civilization back to nature in America tends to be a solitary and ascetic pursuit, having more to do with wilderness than wildness. Johnny Appleseed was very much an *American* Dionysus—innocent and mild. In this he may have helped establish the benign, see-no-evil mood that characterizes the Dionysian strain in American culture, from transcendentalist Concord to the Summer of Love.

∷

"We can hear him now," a woman who had known Chapman reminisced in the 1871 *Harper's* article. "Just as we did that summer day, when we were busy quilting upstairs, and he lay near the door, his voice rising denunciatory and thrilling—strong and loud as the roar of wind and waves, then soft and soothing as the balmy airs that quivered the morning-glory leaves about his gray beard. His was a strange eloquence at times, and he was undoubtedly a man of genius."

Imagine how riveting such a figure must have appeared on the American frontier, this gentle wild man who arrived at your door as if straight from the bosom of nature (wreathed in morning glory leaves, no less). He came bearing ecstatic news from other worlds and, with his apple trees and cider, promising a measure of sweetness in this one. To a pioneer laboring under the brute facts of frontier life, confronting daily the indifferent face of nature, Johnny Appleseed's words and seeds offered release from the long sentence of ordinariness, held out a hope of transcendence.

Under the spell of this otherworldly figure, the world outside one's cabin window suddenly appeared very different, no longer quite so literal or firmly lashed to the here and now. Eyes blazing, Johnny Appleseed would show you how to see the divine in nature, his "strange eloquence" transforming the everyday landscape into a vivid theater of appearances. You could tell yourself this was good Christian doctrine, but truly it was mystical and ecstatic, dwelling more on the all-around sweetness of nature than the singular light of Christ above. And if his words didn't by themselves make the earth flow and the milk and wine and nectar flow like flame, there were then the apple trees he planted, sacramental in their own way, and, perhaps most potent of all, the cider those trees would produce. For one of the wonders of alcohol is that it suffuses the world around us, this cold indifferent planet, with the warm glow of meaning. (Or at least spins that illusion.) This was the gift of sweetness he brought into the country.

:||:

Though Johnny Appleseed may have lacked Dionysus's complementary fierceness, he did deliver in his person a thrilling, scary reminder of the nearness of savagery and the tenuousness of civilization's grip. In both his person and his stories, he temporarily dissolved the stark opposition of wilderness and civilization that organized frontier life. I imagine that pioneers struggling to get by in the wilderness regarded Appleseed as a welcome contrast gainer. However straitened your frontier existence might be, you couldn't gaze on John Chapman without counting your blessings: at least you had leather shoes and a warm hearth, a sociable table and a roof over your head. Your guest's tales of subsisting one winter on butternuts alone, or sharing a bed of leaves with a wolf, would have warmed the draftiest cabin, deepened the savor of the most meager meal. Sometimes the cause of civilization is best served by a hard stare into the soul of its opposite. Some such principle may have underwritten Dionysian revelry in ancient Athens—and the impulse to invite someone like John Chapman into one's home in nineteenth-century Ohio.

:|:

Like Dionysus, John Chapman was an agent of domestication. With every cider orchard he helped plant, the wilderness became that much more hospitable and homelike. (It just happened to be a home he didn't care to live in himself.) But the apple was only one of the many Old World plants John Chapman brought with him into the country; there was the small pharmacopoeia of medicinal herbs, too, and quite a few weeds. I met people in Ohio who still curse Chapman for introducing stinking fennel, a troublesome weed he planted everywhere he went in the belief it could keep a house safe from malaria. (Even today, Ohioans call it "Johnnyweed.") His plantings helped remake the New World landscape in a more familiar image, in the process contributing to an ecological transformation of America the magnitude of which we've just begun to appreciate.

Everyone knows that the settlement of the West depended on the rifle and the ax, yet the seed was no less instrumental in guaranteeing Europeans' success in the New World. (The fact that John Chapman is remembered today along with frontier heroes such as Daniel Boone and Davy Crockett suggests that maybe we knew this before we completely understood it.) The Europeans brought with them to the frontier a kind of portable ecosystem that allowed them to re-create their accustomed way of life—the grasses their livestock needed to thrive, herbs to keep themselves healthy, Old World fruits and flowers to make life comfortable. This biological settlement of the West often went on beneath the notice of the settlers themselves, who brought along weed seeds in the cracks of their boot soles, grass seeds in the feed bags of their horses, and microbes in their blood and gut. (None of these introductions passed beneath the notice of the Native Americans, however.) John Chapman, by planting his millions of seeds, simply went about this work more methodically than most.

In the process of changing the land, Chapman also changed the apple—or rather, made it possible for the apple to change itself. If Americans like Chapman had planted only grafted trees—if Americans had eaten rather than drunk their apples—the apple would not have been able to remake itself and thereby adapt to its new home. It was the seeds, and the cider, that gave the apple the opportunity to discover by trial and error the precise combination of traits required to prosper in the New World. From Chapman's vast planting of nameless cider apple seeds came some of the great American cultivars of the nineteenth century.

Looked at from this angle, planting seeds instead of clones was an extraordinary act of faith in the American land, a vote in favor of the new and unpredictable as against the familiar and European. In this Chapman was making the pioneers' classic wager, betting on the fresh possibilities that might grow from seeds planted in the redemptive American ground. This happens to be nature's wager too, hybridization being one of the ways nature brings newness into the world. John Chapman's millions of seeds and thousands of miles changed the apple, and the apple changed America. No wonder Johnny Appleseed has shaken off the historians and biographers and climbed into our mythology.

⁝⫶⁝

As far as I know, John Chapman never set foot in Geneva, New York, but there is an orchard there where I caught my last and in some ways most vivid glimpse of him. Here on the banks of Seneca Lake, in excellent apple-growing country, a government outfit called the Plant Genetic Resources Unit maintains the world's largest collection of apple trees. Some 2,500 different varieties have been gathered from all over the world and set out here in pairs, as if on a beached botanical ark. The card catalog of this fifty-acre tree archive runs the pomological gamut from Adam's Pearmain, an antique English apple, to the German Zucalmagio. In between a browser will find almost every variety discovered in America since the Roxbury Russet distinguished itself in a cider orchard outside Boston in 1645.

The Geneva orchard is, among other things, a museum of the apple's golden age in America, and a few weeks after my trip to the Midwest, I traveled here, alone, to see what of Johnny Appleseed's legacy I might find in its corridors. At first glance the orchard looks much like any other, the tidy rows of grafted trees advancing like rails to the horizon. But it doesn't take long before you begin to notice the stupendous variety of these trees—in color, leaf, and branching habit—and the metaphor of a library begins to fit: endless shelves of books that are alike only superficially. When I visited, it was late October, and most of the trees were bent with ripe fruit, though many others had already dropped stunning cloaks of red and yellow and green on the ground around them.

I spent the better part of a morning browsing the leafy aisles, tasting all the famous old varieties I'd read about—the Esopus Spitzenberg and Newtown Pippin, the Hawkeye and the Winter Banana. Almost all of these classic varieties were chance seedlings found in exactly the sort of cider orchards John Chapman sponsored, and no doubt there are apples in this orchard that came from the seeds he planted in Pennsylvania, Ohio, and Indiana. There's just no way of knowing which ones they are.

As I worked my way up and down the aisles, consulting a computerized directory that the collection's curator, Phil Forsline, had printed out for me, I concentrated on the varieties listed as "American" and thought about exactly what that meant. By planting so many apples from seed, Americans like Chapman had, willy-nilly, conducted a vast evolutionary experiment, allowing the Old World apple to try out literally millions of new genetic combinations, and by doing so to adapt to the new environment in which the tree now found itself. Every time an apple failed to germinate or thrive in American soil, every time an American winter killed a tree or a freeze in May nipped its buds, an evolutionary vote was cast, and the apples that survived this great winnowing became ever so slightly more American.

A somewhat different kind of vote was then cast by the discriminating orchardist. Whenever a tree growing in the midst of a planting of nameless cider apples somehow distinguished itself—for the hardiness of its constitution, the redness of its skin, the excellence of its flavor—it would promptly be named, grafted, publicized, and multiplied. Through this simultaneous process of natural and cultural selection, the apples took up into themselves the very substance of America—its soil and climate and light, as well as the desires and tastes of its people, and even perhaps a few of the genes of America's native crab apples. In time all these qualities became part and parcel of what an apple in America is.

:‖:

In the years after John Chapman began plying his trade through the Midwest, America witnessed what has sometimes been called the Great Apple Rush. People scoured the countryside for the next champion fruit. The discovery of a Jonathan or Baldwin or Grimes Golden could bring an American fortune and even a measure of fame, and every farmer tended his cider orchard with an eye to the main chance: the apple that would hit it big. "Every wild apple shrub excites our expectations thus," Thoreau wrote, "somewhat as every wild child. It is, perhaps, a prince in disguise. What a lesson to man! . . . Poets and philosophers and statesmen thus spring up in the country pastures, and outlast the hosts of unoriginal men."

The nationwide hunt for pomological genius, the odds of which were commonly held to be eighty thousand to one, brought forth literally hundreds of new varieties, including most of the ones I was now tasting. I can report, however, that not all these children of Chapman are outstanding to eat: many of the apples I sampled that morning were spitters. The Wolf River is particularly memorable in this respect. It had the yellow, wet-sawdust flesh of a particularly tired Red Delicious without even a glint of that apple's beauty.

The sheer profusion of *qualities* that Americans discovered in the apple during its seedling heyday is something to marvel at, especially since so many of those qualities have been lost in the years since. I found apples that tasted like bananas, others like pears. Spicy apples and sticky-sweet ones, apples sprightly as lemons and others rich as nuts. I picked apples that weighed more than a pound, others compact enough to fit in a child's pocket. Here were yellow apples, green apples, spotted apples, russet apples, striped apples, purple apples, even a near-blue apple. There were apples that looked prepolished and apples that wore a dusty bloom on their cheeks. Some of these apples had qualities that were completely lost on me but had meant the world to people once: apples that tasted sweeter in March than October, apples that made especially good cider or preserves or butter, apples that held their own in storage for half a year, apples that ripened gradually to avoid a surfeit or all at once to simplify the harvest, apples with long stem or short, thin skin or thick, apples that tasted sublime only in Virginia and others that needed a hard New England frost to reach perfection, apples that reddened in August, others that held off till winter, even apples that could sit at the bottom of a barrel for the six weeks it took a ship to get to Europe, then emerge bright and crisp enough to command a top price in London.

And the names these apples had! Names that reek of the American nineteenth century, its suspender-popping local boosterism, its shameless Barnum-and-Bailey hype, its quirky, unfocus-grouped individuality. There were the names that set out to describe, often with the help of a well-picked metaphor: the green-as-a-bottle Bottle Greening, the Sheepnose, the Oxheart, the Yellow Bellflower, the Black Gilliflower, the Twenty-Ounce Pippin. There were the names that puffed with hometown pride, like the Westfield Seek-No-Further, the Hubbardston Nonesuch, the Rhode Island Greening, the Albemarle Pippin (though the very same pippin was known as the Newtown nearer to Newtown,

New York), the York Imperial, the Kentucky Red Streak, the Long Stem of Pennsylvania, the Ladies Favorite of Tennessee, the King of Tompkins County, the Peach of Kentucky, and the American Nonpareille. There were names that gave credit where credit was due (or so we assume): the Baldwin, the Macintosh, the Jonathan, McAfee's Red, Norton's Melon, Moyer's Prize, Metzger's Calville, Kirke's Golden Reinette, Kelly's White, and Walker's Beauty. And then there were the names that denoted an apple's specialty, like Wismer's Dessert, Jacob's Sweet Winter, the Early Harvest and Cider Apple, the Clothes-Yard Apple, the Bread and Cheese, Cornell's Savewell and Putnam's Savewell, Paradise Winter, Payne's Late Keeper, and Hay's Winter Wine.

How many other fruits do we call by their Christian names? True, there are a handful of monikered pears and a famous peach or two, but no other fruit in history has produced so many household names—so many celebrities!—as the nineteenth-century apples planted by Chapman and his ilk. Like sports franchises or politicians, each had its contingent of supporters, including a few diehards who could direct you to the semisacred spot on which that apple had first stood (the site often marked with a monument) and recite its biography, the often astounding story of how its genius was first discovered purely by chance, nearly overlooked, and then given its rightful due.

There was the story about the surveyor who stumbled on the Baldwin growing by a Boston canal, or the one about the farmer who noticed the neighborhood boys drawn each winter to the falls around a certain tree that turned out to be the York Imperial, the "king of keepers." And then there was the stubborn, possibly miraculous seedling that kept coming up in between the rows of Jesse Hiatt's orchard in Peru, Iowa, mowing after mowing, until the Quaker farmer decided it must be a sign. So he let the little tree live and fruit, only to discover its apples were far and away the best he'd ever tasted. Hiatt named it the Hawkeye and in 1893 mailed four of them off to a contest at the Stark Brothers Nurseries in Louisiana, Missouri, where C. M. Stark awarded the Hawkeye first prize and a shiny new name: the Delicious. (Stark, a born marketer, had been carrying that name on a slip of paper in his pocket for years, waiting for just the right apple to come along and claim it.) Alas, Jesse Hiatt's entry card was somehow misplaced amid the hoopla, setting off a frantic yearlong search for what would eventually become the world's most popular apple.

There must have been dozens of apple stories more or less in this vein, rags-to-riches fables about a fruit, linking an exemplary tree to a particular American person and place. The parables brought proof not only that the American ground was "fruitful of excellences," in Henry Ward Beecher's excellent phrase, but that Americans themselves had an eye for the main chance and that in America merit would win out in the end. Somehow, this piece of fruit had became a bright metaphor for the American dream.

But why this particular species? Beecher himself said it was because the apple was "the true democratic fruit." Happy to grow just about anywhere, "whether neglected, abused or abandoned, it is able to take care of itself, and to be fruitful of excellences." The Horatio Alger apple that emerged from a nineteenth-century seedling orchard was also in some sense "self-made," something that can't be said about many other plants. The great rose, for example, is the result of careful breeding, the deliberate crossing of aristocratic parents—"elite lines," in the breeder's parlance. Not so the great apple, which distinguishes itself from "the hosts of unoriginal men" without reference to ancestry or breeding. The American orchard, or at least Johnny Appleseed's orchard, is a blooming, fruiting meritocracy, in which every apple seed roots in the same soil and any seedling has an equal chance at greatness, regardless of origin or patrimony.

Befitting the American success story, the botany of the apple—the fact that the one thing it won't do is come true from seed—meant that its history would be a history of heroic individuals, rather than groups or types or lines. There is, or at least there was, a single Golden Delicious tree, of which every subsequent tree bearing that name has been a grafted clone. The original Golden Delicious stood until the 1950s on a hillside in Clay County, West Virginia, where it lived out its golden years inside a padlocked steel cage wired with a burglar alarm. (The cage setup was a publicity stunt organized by Paul Stark, C.M.'s brother, who bought the tree in 1914 for the then-princely sum of $5,000.) Today a granite monument marks the spot where the original Red Delicious grew, between the rows on Jesse Hiatt's Iowa farm. These were two of the many giants that walked what Andrew Jackson Downing called "the young American orchard."

So what native-plant zealot would dare to challenge the right of such trees to call themselves American now? Their ancestors may have evolved half a world away, but these apples have by now undergone much the same process of acculturation as the people who planted them. In fact, they've gone further than the people ever did, for the apples reshuffled their very genes in order to reinvent themselves for life in the New World.

Several of these Americans have since found homes in distant lands (the Golden Delicious now grows on five continents), but many others thrive in America and nowhere else and in some cases are adapted to life in but a single region. The Jonathan, for example, achieves perfection strictly in the American Midwest (which is somewhat surprising, considering it was discovered in the Hudson Valley). My guess is that the Jonathan would be as out of place in England or Kazakhstan, the native ground of its ancestors, as I would be in Russia, the native ground of my own. The arrow of natural history won't be reversed: by now the Jonathan's as much an American as I am.

∷

The golden age of American apples that John Chapman helped to underwrite lives on in the Geneva orchard—yet just about no place else. In fact, the sole reason for its existence is that these erstwhile giants of the young American orchard, the actual and metaphorical descendants of Appleseed's apple seeds, have been all but killed off by the dominance of a few commercially important apples—that, and a pinched modern idea of what constitutes sweetness. A far more brutal winnowing of the apple's prodigious variability took place around the turn of the century. That's when the temperance movement drove cider underground and cut down the American cider orchard, that wildness preserve and riotous breeding ground of apple originality. Americans began to eat rather than drink their apples, thanks in part to a PR slogan: "An apple a day keeps the doctor away." Around the same time, refrigeration made possible a national market for apples, and the industry got together and decided it would be wise to simplify that market by planting and promoting only a small handful of brand-name varieties. That market had no use for the immense variety of qualities the nineteenth-century apple embodied. Now just two of these qualities counted: beauty and sweetness. Beauty in an apple meant a uniform redness, by and large; russeting now doomed even the tastiest apple.

As for sweetness, the complicated metaphorical resonance of that word had by now been flattened out, mainly by the easy availability of cheap sugar. What had been a complex desire had become a mere craving—a sweet tooth. Sweetness in an apple now meant sugariness, plain and simple. And in a culture of easy sweetness, apples now had to compete with every other kind of sugary snack food in the supermarket; even the touch of acid that gives the apple's sweetness some dimension fell out of favor.* And so the Red and Golden Delicious, which are related only by the marketing genius of the Stark brothers (who named and trade-marked them both) and their exceptional sweetness, came to domi-nate the vast, grafted monoculture that the American orchard has become. Apple breeders, locked in a kind of sweetness arms race with junk food, lean heavily on the genes of these two apples, which can be found in most of the popular apples developed in the last few years, including the Fuji and the Gala. Thousands of apple traits, and the genes that code those traits, have become ex-tinct as the vast flowering of apple diversity that Johnny Apple-seed sponsored has been winnowed down to the small handful of varieties that can pass through the needle's eye of our narrow con-ceptions of sweetness and beauty.

This is why the Geneva orchard is a museum. "Today's com-mercial apples represent only a small fraction of the *Malus* gene pool," Phil Forsline, its curator, told me as we walked to a far cor-ner of the orchard, where there was something unusual he wanted me to see. Forsline is a gangly horticulturist in his fifties with striking Nordic blue eyes and sandy hair starting to gray. "A cen-tury ago there were several thousand different varieties of apples in commerce; now most of the apples we grow have the same five or six parents: Red Delicious, Golden Delicious, Jonathan, Macin-

---

*The Granny Smith, a relatively tart green apple discovered in Australia in 1868 (by a Mrs. Smith), is something of an anomaly, though its survival probably owes to its cooking qualities, color, and virtual indestructibility.

tosh, and Cox's Orange Pippin. Breeders keep going back to the same well, and it's getting shallower."

Forsline has devoted a career to preserving and expanding the apple's genetic diversity. He's convinced that the modern history of the apple—particularly the practice of growing a dwindling handful of cloned varieties in vast orchards—has rendered it less fit as a plant, which is one reason modern apples require more pesticide than any other food crop. Forsline explained why this is so.

In the wild a plant and its pests are continually coevolving, in a dance of resistance and conquest that can have no ultimate victor. But coevolution ceases in an orchard of grafted trees, since they are genetically identical from generation to generation. The problem very simply is that the apple trees no longer reproduce sexually, as they do when they're grown from seed, and sex is nature's way of creating fresh genetic combinations. At the same time the viruses, bacteria, fungi, and insects keep very much at it, reproducing sexually and continuing to evolve until eventually they hit on the precise genetic combination that allows them to overcome whatever resistance the apples may have once possessed. Suddenly total victory is in the pests' sight—unless, that is, people come to the tree's rescue, wielding the tools of modern chemistry.

Put another way, the domestication of the apple has gone too far, to the point where the species' fitness for life in nature (where it still has to live, after all) has been dangerously compromised. Reduced to the handful of genetically identical clones that suit our taste and agricultural practice, the apple has lost the crucial variability—the wildness—that sexual reproduction confers.

"The solution is for us to help the apple evolve artificially," Forsline explained, by introducing fresh genes through breeding. A century and a half after John Chapman and others like him seeded the New World with apples, underwriting the orgy of apple sex that led to the myriad new varieties represented in this orchard, another genetic reshuffling may now be necessary. Which is precisely why it is so important to preserve as many different apple genes as possible.

"It's a question of biodiversity," Forsline said as we walked down the long rows of antique apples, tasting as we talked. I was accustomed to thinking of biodiversity in terms of wild species, but of course the biodiversity of the domestic species on which we depend—and which now depend on us—is no less important. Every time an old apple variety drops out of cultivation, a set of genes—which is to say a set of qualities of taste and color and texture, as well as of hardiness and pest resistance—vanishes from the earth.

The greatest biodiversity of any species is typically found in the place where it first evolved—where nature first experimented with all the possibilities of what an apple, or a potato or peach, could be. In the case of the apple, the "center of diversity," as botanists call such a place, lies in Kazakhstan, and in the last few years Forsline has been working to preserve the wild apple genes that he and his colleagues have gathered in the Kazakh forests. Forsline has made several trips to the area, bringing back thousands of seeds and cuttings that he has planted in two long rows all the way in the back of the Geneva orchard. It was these trees, apples far older and wilder than any planted by Johnny Appleseed, that Forsline wanted to show me.

:|:

It was Nikolai Vavilov, the great Russian botanist, who first identified the wild apple's Eden in the forests around Alma-Ata, in Kazakhstan, in 1929. (This wouldn't have come as news to the locals, however: Alma-Ata means "father of the apple.") "All around the city one could see a vast expanse of wild apples covering the foothills," he wrote. "One could see with his own eyes that this beautiful site was the origin of the cultivated apple." Vavilov eventually fell victim to Stalin's wholesale repudiation of genetics, starving to death in a Leningrad prison in 1943, and his discovery was lost to science until the fall of communism. In 1989, one of Vavilov's last surviving students, a botanist named Aimak Djangaliev, invited a group of American plant scientists to see the wild apples he had been studying, very quietly, during the long years of Soviet rule. Djangaliev was already eighty, and he wanted the Americans' help to save the wild stands of *Malus sieversii* from a wave of real estate development spreading out from Alma-Ata to the surrounding hills.

Forsline and his colleagues were astonished to find entire forests of apples, three-hundred-year-old trees fifty feet tall and as big around as oaks, some of them bearing apples as large and red as modern cultivated varieties. "Even in the towns, apple trees were coming up in the cracks of the sidewalks," he recalled. "You looked at these apples and felt sure you were looking at the ancestor of the Golden Delicious or the Macintosh." Forsline determined to save as much of this germ plasm as possible. He felt certain that somewhere among the wild apples of Kazakhstan could be found genes for disease and pest resistance, as well as apple qualities beyond our imagining. Since the wild apple's survival in the wild was now in doubt, he collected hundreds of thousands of seeds, planted as many as he had space for in Geneva, and then offered the rest to researchers and breeders around the world. "I'll send seeds to anybody who asks, just so long as they promise to plant them, tend to the trees, and then report back someday." The wild apples had found their Johnny Appleseed.

:||:

And then there they were, two extravagantly jumbled rows of the weirdest apples I'd ever laid eyes on. The trees had been crammed in cheek by jowl, and the aisles could barely contain, much less order, the luxuriant riot of foliage and fruit, even though it had been planted only six years before. I'd never seen an orchard of apple seedlings (few people nowadays ever do), though it's hard to imagine another seedling orchard quite so crazed by diversity. Forsline had told me that all the apple genes heretofore brought to America—all the genes floating down the Ohio River alongside John Chapman—represented maybe a tenth of the entire *Malus* genome. Well, here was the rest of it.

No two of these trees looked even remotely alike, not in form or leaf or fruit. Some grew straight for the sun, others trailed along the ground or formed low shrubs or simply petered out, the up-state New York climate not to their liking. I saw apples with leaves like those of linden trees, others shaped like demented forsythia bushes. Maybe a third of the trees were bearing fruit—but strange, strange fruit that looked and tasted like God's first drafts of what an apple could be.

I saw apples with the hue and heft of olives and cherries alongside glowing yellow Ping-Pong balls and dusky purple berries. I saw a whole assortment of baseballs, oblate and conic and perfectly round, some of them bright as infield grass, others dull as wood. And I picked big, shiny red fruits that looked just like apples, of all things, though their taste . . . their taste was something else again. Imagine sinking your teeth into a tart potato or a slightly mushy Brazil nut covered in leather. On first bite some of these apples would start out with high promise on the tongue—Now, *here's* an apple!—only to suddenly veer into a bitterness so profound it makes my stomach rise even in recollection.

To get the taste off my tongue, I made for a more civilized row nearby and picked something edible—a Jonagold, I think it was, a cross of Golden Delicious and Jonathan that is to my thinking one of the great achievements of modern apple breeding. And what an achievement that is, to transform a tart potato into a delight of the human eye and tongue. This whole orchard is a testament to the magic arts of domestication, our knack—our Dionysian knack—for marrying the wildest fruits of nature to the various desires of culture. Yet as the modern apple's story suggests, domestication can be overdone, the human quest to control nature's wildness can go too far. To domesticate another species is to bring it under culture's roof, but when people rely on too few genes for too long, a plant loses its ability to get along on its own, outdoors. Something like that happened to the potato in Ireland in the 1840s, and it may be happening to the apple right now.

What saved the potato from that particular blight was genes for resistance that scientists eventually found in wild potatoes growing in the Andes, the potato's own center of diversity. Yet we live in a world where the wild places wild plants live are dwindling. What happens when the wild potatoes and wild apples are gone? The best technology in the world can't create a new gene or re-create one that's been lost. That's why Phil Forsline has dedicated himself to saving and spreading all manner of apples, good, bad, indifferent, and, above all, wild, before it's too late. And that's why all the other sowers of wild seeds, all those who labor under the sign of John Chapman, are to be prized, even if they do blow it now and again, disseminating along with all their good apples the occasional stinking fennel. In the best of all possible worlds we'd be preserving the wild places themselves—the apple's home in the

Kazakh wilderness, for instance. The next best world, though, is the one that preserves the quality of wildness itself, if only because it is upon wildness—of all things!—that domestication depends. That's news to us, perhaps, though Johnny Appleseed was there a century before the scientists and Dionysus a few millennia before him. But how lucky for us that wildness survives in a seed and can be cultivated—can flourish even in the straight lines and right angles of an orchard. "In wildness is the preservation of the world," Thoreau once wrote; a century later, when many of the wild places are no more, Wendell Berry has proposed this necessary corollary: "In human culture is the preservation of wildness."

:|:

A handful of wild apples came home with me from Geneva, a couple of big red ones that caught my eye and a tiny round one no bigger than an olive. This last oddball sat on my desk for a few weeks, and when it started to wrinkle I sliced it through with a knife and scratched out the pippins—five polished ebony seeds that held inside them unimaginable apple mysteries. Who knows what sort of apple would come of such seeds, or of *their* seeds in turn, after the bees crossed their genes with the genes of the Baldwins and Macs in my garden? Probably not an apple you'd want to eat or even look at. But who can say for sure? It was a ridiculous bet, I'll admit, but I decided to give one of the wild apple seeds a spot in my garden anyway—in honor of John Chapman, I suppose, but also just to see what happens.

Though it may not be realistic to expect a sweet apple ever to come of this wildling, I would be surprised if it didn't add something to my garden—if it didn't in some way make it a sweeter place than it is now. Imagine it, this rank, strangely formed tree growing up in a garden, of all places, applelike, perhaps, yet like no apple ever seen and bearing each fall a harvest of strange, unrecognizable fruits. In the middle of a garden—in the middle of a landscape, that is, expressly designed to answer our desires—what such a tree will mostly bear is witness, to an unreconstructed and necessary wildness.

Wallace Stevens wrote a poem about the power of a simple jar sitting on a hill in Tennessee to transform the surrounding forest. He described how this very ordinary bit of human artifice "took dominion everywhere," ordering the "slovenly wilderness" around it like a light in the darkness. I wonder if a wild tree planted in the middle of an ordered landscape can make the reverse happen, can unstring this taut garden, I mean, and allow the cultivated plants all around it to sound the clear note of their own inborn wildness, now muffled. There can be no civilization without wildness, such a tree would remind us, no sweetness absent its astringent opposite.

This garden of mine is bordered by a dwindling contingent of ancient, twisted Baldwins, planted in the twenties by the farmer who built the place and fermented by him, local legend has it, into the tastiest, most potent applejack in town. If nothing else, my aboriginal Kazakh apple tree, growing up in the midst of these, its named and cultivated descendants, will make those old Baldwins taste sweeter than they do now. And if I ever do get around to making a barrel of cider from my Baldwins, a few of these nameless wild apples should add a sharp and racy note to the drink, a strangeness I'll be looking for, and welcome.

CHAPTER 2

Desire: Beauty

Plant: The Tulip

(*TULIPA*)

:|: :|: :|: :|: :|: :|: :|: :|: :|: :|: :|: :|: :|: :|: :|: :|:

T he tulip was my first flower, or at least the first flower I ever planted, though for a long time afterward I was blind to its hard, glamorous beauty. I was maybe ten at the time, and it wasn't until my forties that I could really look at a tulip again. One reason for the long hiatus—for all those years of missed looking—had something to do with the particular tulips I planted as a kid. They would have to have been Triumphs, the tall, blunt, gaily colored orbs you see (or just as often fail to see) massed in the spring land-scape like so many blobs of pigment on a stick. Like the other canonical flowers—the rose or the peony, say—the tulip has been reinvented every century or so to reflect our shifting ideals of beauty, and for the tulip the story of the twentieth century has mainly been the rise and triumph of all this mass-produced eye candy.

Every fall my parents would buy mesh bags of these bulbs, assortments of twenty-five or fifty to the bag, and pay me a few pennies per bulb to bury them in the pachysandra. Presumably they were after something woodsy and naturalistic, which was why they could entrust tulip planting to a ten-year-old boy, whose haphazard and desultory approach was apt to yield exactly the desired effect. I'd press and twist the bulb planter into the root-congested earth until the heel of my hand whitened into a pillowy blister, keeping careful count as I worked, translating the climbing tally of bulbs into the coin of penny candy or trading cards.

October's investment of effort reliably yielded the interest of spring's first color—or perhaps I should say first important color, since the daffodils came earlier. But yellow, besides being commonplace in spring, barely qualifies as a color to a child; red or purple or pink, *those* were colors, and tulips could incarnate them all. This being the early days of the space program, the sturdy tulip stalks reminded me of rockets poised for launch beneath their fat, parti-colored payloads.

These tulips were definitely flowers for kids. They were the simplest of any to draw, and the straightforward spectrum of colors they came in never failed to toe the Crayola line. Accessible and uncomplicated, these run-of-the-garden-center tulips circa 1965 couldn't have been easier for a child to grasp or to grow. But they were easy to grow out of, too, and by the time I was calling the shots in my own garden, a narrow bed of vegetables pressed up against the foundation of our ranch house, I was done with tulips. I thought of myself as a young farmer now and had no time for anything so frivolous as a flower.

⁖

Three and a half centuries earlier, the tulip, still fairly new to the West, unleashed a brief, collective madness that shook a whole nation and nearly brought its economy to ruin. Never before or since has a flower—*a flower!*—taken a star turn on history's main stage as it did in Holland between 1634 and 1637. All that remains of this episode, a speculative frenzy that sucked people at every level of society into its whorl, is a neologism—"tulipomania"—that's not had to be dusted off in all the centuries since, and a historical puzzle. Why there?—in that stolid, parsimonious, Calvinist nation. Why then?—at a time of general prosperity. And why this particular flower?—cool, scentless, and somewhat aloof, the tulip is one of the least Dionysian of flowers, far more likely to elicit admiration than excite passion.

Though something tells me the Triumphs I planted in my parents' pachysandra differed in some key respects from Semper Augustus. Semper Augustus was the intricately feathered red-and-white tulip one bulb of which changed hands for ten thousand guilders at the height of the mania, a sum that at the time would have bought one of the grandest canal houses in Amsterdam. Semper Augustus is gone from nature, though I have seen paintings of it (the Dutch would commission portraits of venerable tulips they couldn't afford to buy), and beside a Semper Augustus a modern tulip looks like a toy.

These are the two poles I want to travel between in these pages: my boyish view of the pointlessness of flowers and the unreasonable passion for them that the Dutch briefly epitomized. The boy's-eye view has the wintry weight of rationality on its side: all this useless beauty is impossible to justify on cost-benefit grounds. But then, isn't that always how it is with beauty? Overboard as the Dutch would eventually go, the fact is that the rest of us—that is, most of humankind for most of its history—have been in the same irrational boat as the seventeenth-century Dutch: crazy for flowers. So what is this tropism all about, for us *and* for the flowers? How did these organs of plant sex manage to get themselves cross-wired with human ideas of value and status and Eros? And what might our ancient attraction for flowers have to teach us about the deeper mysteries of beauty—what one poet has called "this grace wholly gratuitous"? Is that what it is? Or does beauty have a purpose? The story of the tulip—one of the most beloved of flowers, yet a flower curiously hard to love—seems like a good place to search for answers to such questions. Owing to the nature of its object, this particular search doesn't unfold along a straight line. A beeline is more like it—a *real* beeline, though, one that makes a great many stops along its way.

∷

It is possible to be indifferent to flowers—possible but not very likely. Psychiatrists regard a patient's indifference to flowers as a symptom of clinical depression. It seems that by the time the singular beauty of a flower in bloom can no longer pierce the veil of black or obsessive thoughts in a person's mind, that mind's connection to the sensual world has grown dangerously frayed. Such a condition stands as the polar opposite of tulipomania; "flora-ennui," you might call it. It is a syndrome that afflicts individuals, however, not societies.

To judge from my own experience, boys of a certain age also couldn't care less about flowers, regardless of their mental health. For me, fruits and vegetables were the only things to grow, even those vegetables you couldn't pay me to eat. I approached gardening as a form of alchemy, a quasi-magical system for transforming seeds and soil and water and sunlight into things of value, and as long as you couldn't grow toys or LPs, that more or less meant groceries. (I operated a modest farm stand, patronized exclusively by my mother.) To me then (even now), beauty was the breath-catching sight of a glossy bell pepper hanging like a Christmas ornament, or a watermelon nested in a tangle of vines. (Later, briefly, I felt the same way about the five-fingered leaves of a marijuana plant, but that's a special case.) Flowers were all right if you had the space, but what was the point? The flowers I welcomed into my garden were precisely the ones that *had* a point, that foretold the fruit to come: the pretty white-and-yellow button of a strawberry blossom that soon would swell and redden, the ungainly yellow trumpet that heralded the zucchini's coming. Teleological flowers, you might call them.

The other kind, flowers for flowers' sake, seemed to me the flimsiest of things, barely a step up from leaves, which I also deemed of little value; neither ever achieved the sheer existential heft of a tomato or cucumber. The only time I liked tulips was right before they opened, when the flower still formed a closed capsule that resembled some sort of marvelous, weighted fruit. But the day the petals flexed, the mystery drained out of them, leaving behind what to me seemed a weak, papery insubstantiality.

But then, I was ten. What did I know about beauty?

⁝⫶

Aside from certain unimaginative boys, the clinically depressed, and one other exception I will get to, the beauty of flowers has been taken for granted by people for as long as people have been leaving records of what they considered beautiful. Among the treasures the Egyptians made sure the dead had with them on their journey into eternity were the blossoms of flowers, several of which have been found in the pyramids, miraculously preserved. The equation of flowers and beauty was apparently made by all the great civilizations of antiquity, though some—notably the

Jews and early Christians—set themselves against the celebration and use of flowers. But it wasn't out of blindness to their beauty that Jews and Christians discouraged flowers; to the contrary, devotion to flowers posed a challenge to monotheism, was a bright ember of pagan nature worship that needed to be smothered. Incredibly, there were no flowers in Eden—or, more likely, the flowers were weeded out of Eden when Genesis was written down.

This world-historical consensus about the beauty of flowers, which seems so right and uncontroversial to us, is remarkable when you consider that there are relatively few things in nature whose beauty people haven't had to invent. Sunrise, the plumage of birds, the human face and form, and flowers: there may be a few more, but not many. Mountains were ugly until just a few centuries ago ("warts on the earth," Donne had called them, in an echo of the general consensus); forests were the "hideous" haunts of Satan until the Romantics rehabilitated them. Flowers have had their poets too, but they never needed them in quite the same way.

According to Jack Goody, an English anthropologist who has studied the role of flowers in most of the world's cultures—East and West, past and present—the love of flowers is almost, but not quite, universal. The "not quite" refers to Africa, where, Goody writes in *The Culture of Flowers*, flowers play almost no part in religious observance or everyday social ritual. (The exceptions are those parts of Africa that came into early contact with other civilizations—the Islamic north, for example.) Africans seldom grow domesticated flowers, and flower imagery seldom shows up in African art or religion. Apparently when Africans speak or write about flowers, it is usually with an eye to the promise of fruit rather than the thing itself.

Goody offers two possible explanations for the absence of a

culture of flowers in Africa, one economic, the other ecological. The economic explanation is that people can't afford to pay attention to flowers until they have enough to eat; a well-developed culture of flowers is a luxury that most of Africa historically has not been able to support. The other explanation is that the ecology of Africa doesn't offer a lot of flowers, or at least not a lot of showy ones. Relatively few of the world's domesticated flowers have come from Africa, and the range of flower species on the continent is nowhere near as extensive as it is in, say, Asia or even North America. What flowers one does encounter on the savanna, for example, tend to bloom briefly and then vanish for the duration of the dry season.

I'm not sure exactly what to make of the African case, and neither is Goody. Could it mean that the beauty of flowers is in fact in the eye of the beholder—is something people have constructed, like the sublimity of mountains or the spiritual lift we feel in a forest? If so, why did so many different peoples invent it in so many different times and places? More likely, the African case is simply the exception that proves the rule. As Goody points out, Africans quickly adopted a culture of flowers wherever others introduced it. Maybe the love of flowers is a predilection all people share, but it's one that cannot itself flower until conditions are ripe—until there are lots of flowers around and enough leisure to stop and smell them.

:||:

Let's say we *are* born with such a predisposition—that humans, like bees, are drawn instinctively to flowers. It's obvious what good it does bees to be born liking flowers, but what conceivable benefit could such a predilection offer people?

Some evolutionary psychologists have proposed an interesting answer. Their hypothesis can't be proven, at least not until scientists begin to identify genes for human preferences, but it goes like this: Our brains developed under the pressure of natural selection to make us good foragers, which is how humans have spent 99 percent of their time on Earth. The presence of flowers, as even I understood as a boy, is a reliable predictor of future food. People who were drawn to flowers, and who further could distinguish among them and then remember where in the landscape they'd seen them, would be much more successful foragers than people who were blind to their significance. According to the neuroscientist Steven Pinker, who outlines this theory in *How the Mind Works*, natural selection was bound to favor those among our ancestors who noticed flowers and had a gift for botanizing—for recognizing plants, classifying them, and then remembering where they grow. In time the moment of recognition—much like the quickening one feels whenever an object of desire is spotted in the landscape—would become pleasurable, and the signifying thing a thing of beauty.

But wouldn't it make more sense if people were simply hardwired to recognize fruit itself, forget the flowers? Perhaps, but recognizing and recalling flowers helps a forager get to fruit *first*, before the competition. Because I know exactly where on my road the blackberry canes flowered last month, I stand a much better chance of getting to the berries this month before anyone else or any birds do.

I probably should mention at this point that these last speculations are mine, not any scientist's. But I do wonder if it isn't significant that our experience of flowers is so deeply drenched in our sense of time. Maybe there's a good reason we find their fleetingness so piercing, can scarcely look at a flower in bloom without thinking ahead, whether in hope or regret. We might share with certain insects a tropism inclining us toward flowers, but presumably insects can look at a blossom without entertaining thoughts of the past and future—complicated human thoughts that may once have been anything but idle. Flowers have always had important things to teach us about time.

⫶

This is all pure speculation, I know—though speculation itself sometimes seems part and parcel of what a flower is. I'm not sure if they ever asked for it, but flowers have always borne the often absurd weight of our meaning-making, so much so that I'm not prepared to say they *don't* ask for it. Consider, after all, that signifying is precisely what natural selection has designed flowers to do. They were nature's tropes long before we came along.

Natural selection has designed flowers to communicate with other species, deploying an astonishing array of devices—visual, olfactory, and tactile—to get the attention of specific insects and birds and even certain mammals. In order to achieve their objectives, many flowers rely not just on simple chemical signals but on signs, sometimes even on a kind of symbolism. Some plant species go so far as to impersonate other creatures or things in order to secure pollination or, in the case of carnivorous plants, a meal. To entice flies into its inner sanctum (there to be digested by waiting enzymes), the pitcher plant has developed a weirdly striated maroon-and-white flower that is not at all attractive unless you happen to be attracted to decaying meat. (The flower's rancid scent reinforces this effect.)

Ophryus orchids look uncannily like insects, of all things—like bees or flies, depending on the orchid species in question. The Victorians believed this mimicry was intended to scare away insects so the flower could, chastely, pollinate itself. What the Victorians failed to consider was that the Ophryus might resemble an insect precisely in order to attract insects to it. The flower has evolved exactly the right pattern of curves and spots and hairiness to convince certain male insects that it is a female as viewed, tantalizingly, from behind. Botanists call the resultant behavior on the part of the male insect "pseudocopulation"; they call the flower that inspires this behavior the "prostitute orchid." In his frenzy of attempted intercourse, the insect ensures the orchid's pollination. That's because the insect's rising frustration compels him to rush around mounting one blossom after another, effectively disseminating the flower's genes, if not his own.

This stands for that: flowers by their very nature traffic in a kind of metaphor, so that even a meadow of wildflowers brims with meanings not of our making. Move into the garden, however, and the meanings only multiply as the flowers take aim not only at the bee's or the bat's or the butterfly's obscure notions of the good or the beautiful, but at ours as well. Sometime long ago the flower's gift for metaphor crossed with our own, and the offspring of that match, that miraculous symbiosis of desire, are the flowers of the garden.

:|:

In my garden right now it is high summer, the middle of July, and the place is so crowded with flowers, is so busy and multifarious, that it feels more like a city street than a quiet corner of the countryside. At first the scene presents only a daunting confusion of sensory information, a bustle of floral color and scent set to a soundtrack of buzzing insects and rustling leaves, but after a while the individual flowers begin to come into focus. They're the garden's dramatis personae, each of them taking a brief turn on the summer stage, during which it tries its level best to catch our eye. Did I say *our* eye? Well, not *only* ours—for there's also that other audience, the bees and butterflies, moths and wasps and hummingbirds and all the other potential pollinators.

By now the old roses have mostly finished, leaving behind tired shrubs wadded with sad bits of old tissue, but the rugosas and teas are still pumping out color, attracting attention. Tangled up in their petals and seemingly inebriated, the Japanese beetles are dining and humping intently, sometimes three and four of them going at it at once; it's a very Roman scene, and it leaves the blossoms trashed. Farther down the garden path the daylilies lean forward expectantly, like dogs; tiny wasps accept the invitation to climb way up into their throats in search of nectar; afterward the bugs come stumbling out like drunks from a bar. Before they hit the open air, though, they jostle the lily's dainty scoop of stamens, chalking themselves with pollen they'll later dust off on the pistils of some other blossom.*

At the front of the perennial bed the lamb's ears form a low, soft, gray forest of flower spikes that look as though they've been dipped into a vat of bees: the spikes are completely coated, more wing now than petal, and the whole flower is vibrating with the attention. Behind them and high above, the plume poppies throw clouds of tiny white flowers, intricately hairy up close and irresistible to honeybees, who look to be swimming in the air in and

---

*Though many flowers, like the lilies, possess both male and female organs, they go to great lengths to avoid pollinating themselves. That would defeat the floral point, which is the mixing of genes that cross-pollination ensures. A flower can avoid self-pollination chemically (by making its ovule and pollen grain incompatible), architecturally (by arranging stamen and pistil in the flower so as to avoid contact), or temporally (by staggering the times when their stamens produce pollen and their pistils are receptive).

among them. The sweet peas extend themselves seductively on slender stems, but a bee can't gain admittance to their flowers without first prying open their pursed lips; this coy bit of architecture leaves the (erroneous) impression that it is the bee's desire being gratified here, not the pea's.

The bees! The bees will let themselves be lured into the most ridiculous positions, avidly nosing their way like pigs through the thick purple brush of a thistle, rolling around helplessly in a single peony's blond Medusa thatch of stamens—they remind me of Odysseus's crew in thrall to Circe. To my eye the bees appear lost in transports of sexual ecstasy, but of course that's only a projection. It's only a coincidence—*isn't it?*—that this passionate flower-bee embrace that made people think about sex for a thousand years before pollination was understood really *is* about sex. "Flying penises" is what one botanist called bees. But with the rare exception of a flower like the prostitute orchid, for the insects at least it's really not about the sex; to the extent they're penises, they're unwitting penises. Still, the bees certainly do seem besides themselves, and they may well be, but probably on account of the sugary nectars, or maybe one of the designer drugs flowers sometimes deploy in order to drive bees to distraction. Or, who knows, maybe they're just lost in their work.

I've fixed on the bee's-eye view of this scene, but of course the flower's perspective would disclose that in the garden human desire looms just as large. In fact, the place is crowded with species that have evolved expressly to catch *my* eye, often to the detriment of getting themselves pollinated. I'm thinking of all the species that have sacrificed their scent in the interest of grander or doubled or improbably colored blooms, ideals of beauty that probably go unappreciated in the kingdom of the pollinators, a place where the eye is not always sovereign.

For many flowers the great love of their lives now is human-kind. Those daylilies leaning expectantly forward? Their faces are in fact turned toward us, whose favor now ensures their success better than any bug's can. That peony with the salacious pubic stamens? Blame the Chinese for that one: for thousands of years their poets, discerning manifestations of yin and yang in the garden, likened peony blossoms to a woman's sexual organs (and the bee or butterfly to a man's); over time Chinese peonies evolved, by means of artificial selection, to gratify that conceit. Even the perfume of certain Chinese tree peonies is womanly, a scent of flowers tinged with briny sweat; the flowers smell less like perfume out of the bottle than a scent that's spent time on human skin. It may still attract the bees, but by now it's our brain stems the scent is meant to fire.

∷

Making my way through this lit-up landscape, I try to pin down exactly what distinguishes the garden in bloom from an ordinary patch of nature. For starters, the flowering garden is a place you immediately sense is thick with information, thick as a metropolis, in fact. It's an oddly sociable, public sort of place, in which species seem eager to give one another the time of day; they dress up, flirt, flit, visit. By comparison, the surrounding forests and fields are much sleepier boroughs, steadily humming monotonies of green, in which many of the flowers are inconspicuous or short-lived and many of the plants seem to be keeping to their own kind, declining to enlist other species, minding their own business. That business is chiefly photosynthesis, of course, nature's routine factory work; sexual reproduction is going on here too, but with little to show for it: Who ever notices when the conifers release their pollen on the wind, the ferns their minute spores? April through October, every day looks pretty much the same around here. What beauty there is is in large part inadvertent, purposeless, and unadvertised.

Come into the garden, or even the flowering meadow, and the landscape immediately quickens. Hey, what's going on here today? *Something,* senses even the dimmest bee or boy, something special. Call that something the stirrings of beauty. Beauty in nature often shows up in the vicinity of sex—think of the plumage of birds or mating rituals throughout the animal kingdom. "Sexual selection"—that is, evolution's favoring of features that increase a plant's or animal's attractiveness and therefore its reproductive success—is the best explanation we have for the otherwise senseless extravagance of feathers and flowers, maybe also sports cars and bikinis. In nature, at least, the expense of beauty is usually paid for by sex.

There may or may not be a correlation between the beautiful and the good, but there probably is one between beauty and health. (Which, I suppose, in Darwinian terms, *is* the good.) Evolutionary biologists believe that in many creatures beauty is a reliable indicator of health, and therefore a perfectly sensible way to choose one mate over another. Gorgeous plumage, lustrous hair, symmetrical features are "certificates of health," as one scientist puts it, advertisements that a creature carries genes for resistance to parasites* and is not otherwise under stress. A fabulous tail is a metabolic extravagance only the healthy can afford. (In the same way, a fabulous car is a financial extravagance only the successful can afford.) In our own species, too,

---

*Among birds, the species most susceptible to parasites are the ones with the most extravagant plumage—probably because these are the ones that most need to advertise their fitness.

ideals of beauty often correlate with health: when lack of food was what usually killed people, people judged body fat to be a thing of beauty. (Though the current preference for sickly-pale, rail-thin models suggests that culture can override evolutionary imperatives.)

But what about plants, who don't get to choose their mates? Why should the bees, who do the choosing for them, care a fig about plant health? They don't, yet unwittingly they reward it. It's the healthiest flowers that can afford the most extravagant display and sweetest nectar, thereby ensuring the most visits from bees— and therefore the most sex and most offspring. So in a sense, the flowers do choose their mates on the basis of health, using the bees as their proxies.

∷

Before the advent of this arms race of sexual selection—before flowers, before feathers—all nature was the factory. There was beauty there, but it was not beauty by design; what beauty there was was, like that of forests or mountains, strictly in the eye of the beholder.

If you wanted to invent a new myth of the origin of beauty (or at least designed beauty), you could do worse than begin here in the garden, among the flowers. Begin with the petal, where beauty's first principle—contrast with its surroundings—appears, a feat here accomplished with color. The eye, lulled by the all-around green all around, registers the difference and rouses. Bees, once thought to be color-blind, do in fact see color, though they see it differently than we do. Green appears gray, a background hue against which red—which bees perceive as black—stands out most sharply. (Bees can also see at the ultraviolet end of the spectrum, where we're blind; a garden in this light must look like a big-city airport at night, lit up and color-coded to direct circling bees to landing zones of nectar and pollen.)

Bee or boy, our attention is awakened by a petal's color, alerting us to what comes next, which is form or pattern, beauty's second inflection of the given world. Against the background of inchoate green a contrasting color by itself could well be an accident of some kind (a feather, say, or a dying leaf), but the appearance of symmetry is a reliable expression of formal organization—of purpose, even intent. Symmetry is an unmistakable sign that there's relevant information in a place. That's because symmetry is a property shared by a relatively small number of things in the landscape, all of them of keen interest to us. The shortlist of nature's symmetricals includes other creatures, other people (most notably the faces of other people), human artifacts, and plants—but especially flowers. Symmetry is also a sign of health in a creature, since mutations and environmental stresses can easily disturb it. So paying attention to symmetrical things makes good sense: symmetry is usually significant.

The same holds true for bees. How do we know? Because symmetry in a plant is an extravagance (whereas animals who want to move in a straight line can't do without it), and natural selection probably wouldn't go to the trouble if the bees didn't reward the effort. "The colors and shapes of the flowers are a precise record of what bees find attractive," the poet and critic Frederick Turner has written. He goes on to suggest that it "would be a paradoxically anthropocentric mistake to assume that, because bees are more primitive organisms . . . there is nothing in common between our pleasure in flowers and theirs."

But if the pleasure bees and people take in flowers have a common root, standards of floral beauty soon begin to specialize and diverge—and not just bee from boy, but bee from bee as well. For it seems that different kinds of bees are attracted to different kinds of symmetry. Honeybees favor the radial symmetry of daisies and clover and sunflowers, while bumblebees prefer the bilateral symmetry of orchids, peas, and foxgloves.*

Through their colors and symmetries, through these most elemental principles of beauty (that is, contrast and pattern), flowers alert other species to their presence and significance. Walk among them, and you see faces turned toward you (though not only you), beckoning, greeting, informing, promising—*meaning*. Beyond that, matters begin to get complicated, the honeybees developing their own canons of beauty, the bumblebees theirs. And then into this great dance of plants and pollinators step us, compounding the meanings of flowers beyond all reason, turning their sexual organs into tropes of our own (and of so much else), drawing and driving the evolution of flowers toward the extraordinary, freakish, and precarious beauty of a Madame Hardy rose or a Semper Augustus tulip.

∷

There are flowers, and then there are flowers: flowers, I mean, around which whole cultures have sprung up, flowers with an empire's worth of history behind them, flowers whose form and color and scent, whose very genes carry reflections of people's ideas and desires through time like great books. It's a lot to ask of a plant, that it take on the changing colors of human dreams, and this may explain why only a small handful of them have proven themselves supple and willing enough for the task. The rose, obviously, is one such flower; the peony, particularly in the East, is another. The

---

*Whatever the case, the more perfect the symmetry, the healthier—and therefore sweeter—the flower.

orchid certainly qualifies. And then there is the tulip. Arguably there are a couple more (perhaps the lily?), but these few have long been our canonical flowers, the Shakespeares, Miltons, and Tolstoys of the plant world, voluminous and protean, the select company of flowers that have survived the vicissitudes of fashion to make themselves sovereign and unignorable.

So what sets these flowers apart from the run of charming daisies and pinks and carnations, not to mention the legions of pretty wildflowers? Perhaps more than anything else, it is their multifariousness. Some perfectly good flowers simply are what they are, singular and, if not completely fixed in their identity, capable of ringing only a few simple changes on it: hue, say, or petal count. Prod it all you want, select and cross and reengineer it, but there's only so much a coneflower or a lotus is ever going to do. Fashion is apt to pick up such a flower for a time and then drop it—think of the pink, or gillyflower, in Shakespeare's day or the hyacinth in Queen Victoria's—since it won't let itself be remade in some new image once its first one is passé.

By contrast, the rose, the orchid, and the tulip are capable of prodigies, reinventing themselves again and again to suit every change in the aesthetic or political weather. The rose, flung open and ravishing in Elizabethan times, obligingly buttoned herself up and turned prim for the Victorians. When the Dutch decided the paragon of floral beauty was a marbleized swirl of vividly contrasting colors, the petals of their tulips became extravagantly "feathered" and "flamed." But then, when the English went in big for "carpet bedding" in the nineteenth century, the tulips duly allowed themselves to be turned into a paint box filled with the brightest, fattest dabs of pure pigment, suitable for massing. These are the sorts of flowers that bear our oddest notions gladly. Of course, their willingness to take part in the moving game of human culture has proven a brilliant strategy for their success, for there are a lot more roses and tulips around today, in a lot more places, than there were before people took an interest in them. For a flower the path to world domination passes through humanity's ever-shifting ideals of beauty.

:|:

It isn't automatically obvious that the tulip belongs in this august company of flowers, probably because, in its modern incarnation, the tulip is such a simple, one-dimensional flower, and its rich history of being so much more than that has largely been lost. Compared to the rose or the peony, flowers whose historical forms survive alongside their modern incarnations (both because the plants are so long-lived and because they can be cloned indefinitely), the only way we have any idea what made a tulip beautiful in Turkish or Dutch or French eyes is through those people's paintings and botanical illustrations. That's because a tulip that falls out of favor soon goes extinct, since the bulbs don't reliably come back every year. In general a strain won't last unless it is regularly replanted, so the chain of genetic continuity can be broken in a generation. Even when people do continue to plant a particular tulip, the vigor of that variety (which is propagated by removing and planting the bulb's "offsets," the little, genetically identical bulblets that form at its base) eventually fades until it must be abandoned. Breeders today are busily seeking a new black tulip because they know the current standard-bearer—Queen of Night—is probably on her way out. Tulips, in other words, are mortal.

:|:

No tulip appears in the flower-crowded borders of medieval tapestries, nor is the flower ever mentioned in the early "herbals"— the Old World encyclopedias of the world's known plants and their uses. The fierceness of the passion that the tulip unleashed in Holland in the seventeenth century (and to a lesser extent in France and England) may have had something to do with the flower's novelty in the West and the suddenness of its appearance. It is the youngest of our canonical flowers, the rose being the oldest.

Ogier Ghislain de Busbecq, ambassador of the Austrian Hapsburgs to the court of Süleyman the Magnificent in Constantinople, claimed to have introduced the tulip to Europe, sending a consignment of bulbs west from Constantinople soon after he arrived there in 1554. (The word *tulip* is a corruption of the Turkish word for "turban.") The fact that the tulip's first official trip west took it from one court to another—that it was a flower favored by royalty—may also have contributed to its quick ascendancy, for court fashions have always been especially catching.

The tulip's is not a case where a plant had to travel the world before its virtues could be recognized at home: by the time of Busbecq's consignment, the tulip already had its own cult of admirers in the East, who had taken the flower a considerable distance from its form in the wild. There, it typically appears as a short, pretty, cheerful flower, a frank, open-faced, six-petaled star, often with a dramatic splotch of contrasting color at the base. Species tulips in Turkey typically come in red, less commonly in white or yellow. The Ottoman Turks had discovered that these wild tulips were great changelings, freely hybridizing (though it takes seven years before a tulip grown from seed flowers and shows its new colors) but also subject to mutations that produced spontaneous and wondrous changes in form and color. The tulip's mutability was taken as a sign that nature cherished this flower above all others. In his 1597 herbal, John Gerard says of the tulip that "nature seems to plaie more with this flower, than with any other that I do know."

The tulip's genetic variability has in fact given nature—or, more precisely, natural selection—a great deal to play with. From among the chance mutations thrown out by a flower, nature preserves the rare ones that confer some advantage—brighter color, more perfect symmetry, whatever. For millions of years such features were selected, in effect, by the tulip's pollinators—that is, insects—until the Turks came along and began to cast their own votes. (The Turks did not learn to make deliberate crosses till the 1600s; the novel tulips they prized were said simply to have "occurred.") Darwin called such a process artificial, as opposed to natural, selection, but from the flower's point of view, this is a distinction without a difference: individual plants in which a trait desired by either bees or Turks occurred wound up with more offspring. Though we self-importantly regard domestication as something people have done to plants, it is at the same time a strategy by which the plants have exploited us and our desires—even our most idiosyncratic notions of beauty—to advance their own interests. Depending on the environment in which a species finds itself, different adaptations will avail. Mutations that nature would have rejected out of hand in the wild sometimes prove to be brilliant adaptations in an environment that's been shaped by human desire.

In the environment of the Ottoman Empire the best way for a tulip to get ahead was to have absurdly long petals drawn to a point fine as a needle. In drawings, paintings, and ceramics (the only place the Turks' ideal of tulip beauty survives; the human environment is an unstable one), these elongated blooms look as though they'd been stretched to the limit by a glassblower. The metaphor of choice for this form of tulip petal was the dagger. A successful Ottoman tulip also had to be pure in color and have smooth-edged petals held closely enough together to hide the anthers within, and it could never be "doubled"—have a super-abundance of petals, in the way of a hybrid rose. Though these last traits are not uncommon in species tulips, attenuated petals are virtually unknown in the wild, which suggests that the Ottoman ideal of tulip beauty—elegant, sharp, and masculine—was freakish and hard-won and conferred no advantage in nature. (Very often traits that commend plants and animals to people render them less fit for life in the wild.) Beyond a certain point the Ottoman and insect ideals of tulip beauty no longer coincided.

For a time in the eighteenth century the bulbs of tulips that matched the Turkish ideal traded in Constantinople for quantities of gold. This was during the reign of Sultan Ahmed III, from 1703 to 1730, a period known to Turkish historians as the *lale devri*, or Tulip Era. The sultan was ruled by his passion for the flower, so much so that he imported bulbs by the millions from Holland, where the Dutch, after the passing of their own tulipomania, had become masters of large-scale bulb production. The extravagance of the sultan's annual tulip festivals ultimately proved his downfall; the conspicuous waste of national treasure helped fire the revolt that ended his rule.

Each spring for a period of weeks the imperial gardens were filled with prize tulips (Turkish, Dutch, Iranian), all of them shown to their best advantage. Tulips whose petals had flexed too wide were held shut with fine threads hand-tied. Most of the bulbs had been grown in place, but these were supplemented by thousands of cut stems held in glass bottles; the scale of the display was further compounded by mirrors placed strategically around the garden. Each variety was marked with a label made from silver filigree. In place of every fourth flower a candle, its wick trimmed to tulip height, was set into the ground. Songbirds in gilded cages supplied the music, and hundreds of giant tortoises carrying candles on their backs lumbered through the gardens, further illuminating the display. All the guests were required to dress in colors that flattered those of the tulips. At the appointed moment a cannon sounded, the doors to the harem were flung open, and the sultan's mistresses stepped into the garden led by eunuchs bearing torches. The whole scene was repeated every night for as long as the tulips were in bloom, for as long as Sultan Ahmed managed to cling to his throne.

:||:

A theft lies behind the rise of the tulip in Holland. One of the recipients of the first tulips to arrive in Europe was Carolus Clusius, a cosmopolitan plantsman who played a seminal role in the distribution of newly discovered plants through Europe. Bulbs were his specialty, and Clusius is credited with the introduction, or spread, of fritillarias, irises, hyacinths, anemones, ranunculi, narcissi, and lilies. The tulips came into Clusius's hands because he was director of the Imperial Botanical Garden in Vienna. When he moved to Leiden to establish a new physic garden in 1593, he took some of the bulbs with him.

According to Anna Pavord's history of the tulip, the flower was already growing, with little fanfare, in at least one Leiden garden by the time of Clusius's arrival. But Clusius was so ostentatiously possessive of his rare tulips that he made the Dutch covet them, with disastrous consequences for his collection. In the words of one contemporary account, "No one could procure them, not even for money. [So] plans were made by which the best and most of his plants were stolen by night whereupon he lost courage and the desire to continue their cultivation; but those who had stolen the tulips lost no time in increasing them by sowing the seeds, and by this means the seventeen provinces were well stocked."

Two things about this story are noteworthy. The first is that the stolen tulips were propagated by seed. Tulips, like apples, do not come true from seed—their offspring bear little resemblance to their parents. What this means is that, given the flower's inherent variability, the seventeen provinces of Holland would have been "stocked" with an extraordinary array of differently shaped and colored tulips. This promiscuous seeding of tulips may well have been the source of much of the astounding variety the Dutch managed to coax from the flower, a botanical treasure that became a point of national pride in the seventeenth century. Holland's tulips were mentioned in the same breath as its invincible navy and unparalleled republican liberties.

The second noteworthy point about the story is that it puts a theft at the source of Holland's long, illustrious, and ignominious relationship with the tulip. (This was not the first or last time a theft attended the appearance of a new plant; the potato might never have prospered in France if not for a similar theft from the royal gardens of Louis XVI.) Very often in myth a theft, and its consequence of shame, lies at the root of a human achievement—think of Prometheus's theft of fire from the sun or Eve's tasting of the fruit of knowledge. Shame seems to be the going price of achievement, particularly the achievement of knowledge or beauty. For the Dutch, at least, shame has shadowed the tulip's story from the start, though fainter manifestations of the same shadow are probably never far from the culture of flowers. It's there in the wastefulness and extravagance we often associate with flowers, in the sensual pleasure we take in them, in our satisfaction at forcing them beyond their natural forms and colors and blooming times, even in the tiny pang that can accompany the petty theft of a flower that's been cut and brought indoors.

:|:

The modern tulip has become such a cheap and ubiquitous commodity that it's hard for us to recover a sense of the glamour that once surrounded the flower. That glamour surely had something to do with its roots in the Orient—Anna Pavord speaks of the "intoxicating aura of the infidels" that surrounded the tulip. There was, too, the preciousness of the early tulips, the supply of which could be increased only very slowly through offsets, a quirk of biology that kept supply well behind demand. In France in 1608, a miller exchanged his mill for a bulb of Mère Brune. Around the same time a bridegroom accepted a single tulip as the whole of his dowry—happily, we are told; the variety became known as "Mariage de ma fille."

Yet tulipomania in France and England never reached the pitch it would in Holland. How can the mad embrace of these particular people and this particular flower be explained?

For good reason, the Dutch have never been content to accept nature as they found it. Lacking in conventional charms and variety, the landscape of the Low Countries is spectacularly flat, monotonous, and swampy. "An universall quagmire" is how one Englishman described the place; "the buttock of the world." What beauty there is in the Netherlands is largely the result of human effort: the dikes and canals built to drain the land, the windmills erected to interrupt the unbroken sweep of wind across it. In his famous essay on tulipomania, "The Bitter Smell of Tulips," the poet Zbigniew Herbert suggests that the "monotony of the Dutch landscape gave rise to dreams of multifarious, colorful, and unusual flora."

Such dreams could be indulged as never before in seventeenth-century Holland, as Dutch traders and plant explorers returned home with a parade of exotic new plant species. Botany became a national pastime, followed as closely and avidly as we follow sports today. This was a nation, and a time, in which a botanical treatise could become a best-seller and a plantsman like Clusius a celebrity.

Land in Holland being so scarce and expensive, Dutch gardens were miniatures, measured in square feet rather than acres and frequently augmented with mirrors. The Dutch thought of their gardens as jewel boxes, and in such a space even a single flower—and especially one as erect, singular, and strikingly colored as a tulip—could make a powerful statement.

To make such statements—about one's sophistication, about one's wealth—has always been one of the reasons people plant gardens. In the seventeenth century the Dutch were the richest people in Europe and, as the historian Simon Schama shows in *The Embarrassment of Riches,* their Calvinist faith did not keep them from indulging in the pleasures of conspicuous display. The exoticism and expense of tulips certainly recommended them for this purpose, but so did the fact that, among flowers, the tulip is one of the most extravagantly useless. Up until the Renaissance, most of the flowers in cultivation had been useful as well as beautiful; they were sources of medicine, perfume, or even food. In the West flowers have often come under attack from various Puritans, and what has always saved them has been their practical uses. It was utility, not beauty, that earned the rose and lily, the peony and all the rest a spot in the gardens of monks and Shakers and colonial Americans who would otherwise have had nothing to do with them.

When the tulip first arrived in Europe, people set about fashioning some utilitarian purpose for it. The Germans boiled and sugared the bulbs and, unconvincingly, declared them a delicacy; the English tried serving them up with oil and vinegar. Pharmacists proposed the tulip as a remedy for flatulence. None of these uses caught on, however. "The tulip remained itself," Herbert writes, "the poetry of Nature to which vulgar utilitarianism is foreign." The tulip was a thing of beauty, no more, no less.

If the tulip's useless beauty suited the Dutch taste for display, it also meshed with the age's humanism, which was striving to put some breathing space between art and religion. Unlike the rose or the lily, say, the tulip had not yet been enlisted as a Christian symbol (though tulipomania would eventually change that); to paint a vase of tulips was to delve into the wonders of nature rather than into the storehouse of iconography.

I also think the particular character of the tulip's beauty made it a good match for the Dutch temperament. Generally bereft of scent, the tulip is the coolest of floral characters. In fact, the Dutch counted the tulip's lack of scent as a virtue, a proof of the flower's chasteness and moderation. Petals curving inward to hide its sexual organs, the tulip is an introvert among flowers. It is also somewhat aloof—one bloom per stem, one stem per plant. "The tulip allows us to admire it," Herbert observes, "but does not awaken violent emotions, desire, jealousy or erotic fevers."

None of these qualities would seem to portend the frenzy to come. But as it would happen, the outward composure of Dutchman and tulip alike held sleeping within it something else.

⁞

One crucial element of the beauty of the tulip that intoxicated the Dutch, the Turks, the French, and the English has been lost to us. To them the tulip was a magic flower because it was prone to spontaneous and brilliant eruptions of color. In a planting of a hundred tulips, one of them might be so possessed, opening to reveal the white or yellow ground of its petals painted, as if by the finest brush and steadiest hand, with intricate feathers or flames of a vividly contrasting hue. When this happened, the tulip was said to have "broken," and if a tulip broke in a particularly striking manner—if the flames of the applied color reached clear to the petal's lip, say, and its pigment was brilliant and pure and its pattern symmetrical—the owner of that bulb had won the lottery. For the offsets of that bulb would inherit its pattern and hues and command a fantastic price. The fact that broken tulips for some unknown reason produced fewer and smaller offsets than ordinary tulips drove their prices still higher. Semper Augustus was the most famous such break.

The closest we have to a broken tulip today is the group known as the Rembrandts—so named because Rembrandt painted some of the most admired breaks of his time. But these latter-day tulips, with their heavy patterning of one or more contrasting colors, look clumsy by comparison, as if painted in haste with a thick brush. To judge from the paintings we have of the originals, the petals of broken tulips could be as fine and intricate as marbleized papers, the extravagant swirls of color somehow managing to seem both bold and delicate at once. In the most striking examples—such as the fiery carmine that Semper Augustus splashed on its pure white ground—the outbreak of color juxtaposed with the orderly, linear form of the tulip could be breathtaking, with the leaping, wayward patterns just barely contained by the petal's edge.

Anna Pavord recounts the extraordinary lengths to which Dutch growers would go to make their tulips break, sometimes borrowing their techniques from alchemists, who faced what must have seemed a comparable challenge. Over the earth above a bed planted with white tulips, gardeners would liberally sprinkle paint powders of the desired hue, on the theory that rainwater would wash the color down to the roots, where it would be taken up by the bulb. Charlatans sold recipes believed to produce the magic color breaks; pigeon droppings were thought to be an effective agent, as was plaster dust taken from the walls of old houses. Unlike the alchemists, whose attempts to change base metals into gold reliably failed, now and then the would-be tulip changers would be rewarded with a good break, inspiring everybody to redouble their efforts.

What the Dutch could not have known was that a virus was responsible for the magic of the broken tulip, a fact that, as soon as it was discovered, doomed the beauty it had made possible. The color of a tulip actually consists of two pigments working in concert—a base color that is always yellow or white and a second, laid-on color called an anthocyanin; the mix of these two hues determines the unitary color we see. The virus works by partially and irregularly suppressing the anthocyanin, thereby allowing a portion of the underlying color to show through. It wasn't until the 1920s, after the invention of the electron microscope, that scientists discovered the virus was being spread from tulip to tulip by *Myzus persicae,* the peach potato aphid. Peach trees were a common feature of seventeenth-century gardens.

By the 1920s the Dutch regarded their tulips as commodities to trade rather than jewels to display, and since the virus weakened the bulbs it infected (the reason the offsets of broken tulips were so small and few in number), Dutch growers set about ridding their fields of the infection. Color breaks, when they did occur, were promptly destroyed, and a certain peculiar manifestation of natural beauty abruptly lost its claim on human affection.

I can't help thinking that the virus was supplying something the tulip needed, just the touch of abandon the flower's chilly formality called for. Maybe that's why the broken tulip became such a treasure in seventeenth-century Holland: the wayward color loosed on a tulip by a good break perfected the flower, even as the virus responsible set about destroying it.

.:|:.

On its face the story of the virus and the tulip would seem to throw a wrench into any evolutionary understanding of beauty. What possible good could it do a flower for an infection that decreases its fitness to enhance its appeal to people? I suppose a case could be made that the virus, by adding fuel to the frenzy of tulipomania, led to the planting of many more tulips in the hope of finding more breaks. But the fact remains that, because of people's idiosyncratic notion of tulip beauty, for several hundred years tulips were selected for a trait that would sicken and eventually kill them.

This would seem to represent a perversion of natural selection, a violation of the laws of nature. And so it is—considered from the vantage point of the tulip. But what if the question is considered instead from the vantage point of the virus? The rule of law is restored. What the virus did was to insinuate itself into the relationship between people and flowers, in effect exploiting human ideas of tulip beauty in order to advance its own selfish purposes. (Which, if you think about it, is not so different from what humans did when they elbowed into the old relationship of bees and flowers.) The more beautiful the breaks produced by the infection, the greater the number of infected plants in Dutch gardens and the more total virus in circulation. What a trick! As a survival strategy, the virus's scheme was brilliant, at least as long as people didn't figure out what was going on. For where else in nature has a disease rendered a living thing more lovely? And not just lovely, but lovely in a previously unimagined way, for the virus created an entirely new way for a tulip to be beautiful, at least in our eyes. The virus altered the eye of the beholder. That this change came at the expense of the beheld suggests that beauty in nature does not necessarily bespeak health, nor necessarily redound to the benefit of the beautiful.

∷

The transformation of the tulip from a jewel-box flower to a (virus-free) commodity has made the tulip oddly hard to see. Massed in the landscape, tulips register on us mostly as instances of pure color; they could almost be lollipops or lipsticks in the landscape. At least this is how they used to register on me—as eye candy, pleasurable enough but weightless. I am not by nature a great noticer, and for all the years between the time when my parents paid me to plant tulips in our yard and the spring of this writing, the beauty of tulips—their specific beauty—was lost on me. But I don't think the problem is unique to me.

"Beauty always takes place in the particular," the critic Elaine Scarry has written, "and if there are no particulars, the chances of seeing it go down." In a sense, particular tulips are hard to come by—because they are so cheap and ubiquitous, that's partly why, but also because their form and color are, more than those of most flowers, peculiarly abstract. Far more than a rose, say, or a peony, an actual, specific tulip closely resembles our preconceived idea of a tulip. By now the tulip's parabolic curves are as deeply etched into consciousness as a Coke bottle's; with a fidelity that is remarkable (and that is far more typical of a commodity than a thing in nature), the tulips one meets in the world match the tulips resident in one's head. In color, too, tulips are so uniform and faithful (like paint chips) to whatever shade they profess to be that we quickly take it in—this *idea* of yellow or red or white— and then move on to consume the next visual treat. Tulips are so tuliplike, so platonically themselves, that they skate past our regard like models on a runway.

    :|:

One way to begin to slow down and recover the particular beauty of a tulip, I discovered this spring, is to bring one indoors and look at it individually. This, I think, may be even more helpful than planting older or more exotic varieties, for I suspect that even some of the Triumphs and Darwins sold in the mass-market mesh bags would, if cut and brought indoors and then really looked at, also hold the power to astonish. It is no accident that botanical illustrators and photographers have so often brought their scrupulous eye to bear on this particular flower: it rewards that particular gaze like no other.

I eventually want to bring that gaze briefly to bear on a single tulip—the Queen of Night sitting before me on my desk this late-May morning. Queen of Night is as close to black as a flower gets, though in fact it is a dark and glossy maroonish purple. Its hue is so dark, however, that it appears to draw more light into itself than it reflects, a kind of floral black hole. In the garden, depending on the angle of the sun, the blossoms of a Queen of Night may read as positive or negative space, as flowers or shadows of a flower.

This particular effect was prized by the Dutch, and the quest for a truly black tulip—a quest that has gone on for at least four hundred years and goes on still—became one of the more intriguing subplots of tulipomania. Alexandre Dumas *père* wrote a whole novel—*The Black Tulip*—about a competition in seventeenth-century Holland to grow the first truly black tulip; the greed and intrigue inspired by the contest (in the novel the Horticultural Society had put up a prize of 100,000 guilders) destroyed three lives. By the time the "miraculous tulip" appears, Cornelius, the man who bred it, is in jail, wrongly imprisoned on a tip by his neighbor, who has claimed the prize flower as his own. Cornelius glimpses the culmination of his life's work through the bars of his cell: "The tulip was beautiful, splendid, magnificent; its stem was more than eighteen inches high. It rose from out of four green leaves, which were as smooth and straight as iron lance heads; the whole of the flower was as black and shining as jet."

But why a *black* tulip? Perhaps because the color black is so rare in nature (or at least, in living nature), and tulipomania was nothing if not a vast and precarious edifice poised on the finest points of botanical rarity. Black also carries connotations of evil, and the mania would later come to be seen as a morality tale about worldly temptation, in which a whole people succumbed, ruinously, to not one but an entire bouquet of deadly sins. At the same time, black, like white, is a blankness onto which any and all desire (or fear) may be projected. For Dumas the black tulip was a synecdoche for tulipomania itself, an indifferent and arbitrary mirror in which a perverse consensus of meaning and value came briefly and disastrously into focus.

A second story is told, this one possibly true, about a black tulip discovered by a poor shoemaker at the height of the madness. In the version that Zbigniew Herbert tells, five gentlemen from the union of florists in Haarlem, all dressed in black, pay a visit to the shoemaker, professing to do him a good turn by offering to buy his tulip bulb. The shoemaker, sensing their avarice, begins to bargain in earnest, and after much haggling the two parties settle on a price for the bulb: 1,500 florins, a sum that to the shoemaker is a windfall. The bulb changes hands.

"Now something unexpected happened," Herbert writes, "something that in drama is called a turning point." The florists throw the precious bulb to the ground and stomp it to a pulp.

" 'You idiot!' they shouted at the stupefied shoe patcher, 'we also have a bulb of the black tulip. Besides us, no one else in the world! No king, no emperor or sultan. If you had asked ten thousand florins for your bulb and a couple of horses on top of it, we would have paid without a word. And remember this. Good fortune won't smile on you a second time in your entire life, because you are a blockhead.' " The shoemaker, devastated, staggers to his bed in the attic and dies.

Herbert's view of the tulipomania is itself unremittingly black. To him the Dutch frenzy had nothing whatever to do with beauty, only with the consuming evil of the fixed idea, a phenomenon that can, at any time, destroy the "sanctuaries of reason" on which civilization depends. Herbert's tulipomania is a parable of utopianism, specifically of communism. It is true that, after a certain point, the flowers themselves became irrelevant—a time came when crushing a particular tulip bulb, or holding a paper "futures contract" for another still in the ground, conferred greater wealth than the most beautiful blossom ever beheld.

Still, it's important to remember that what ended in Holland in madness had begun with the desire for beauty in a place where, it seemed to many, beauty was in comparatively short supply. This was also a country, remember, where everyone, regardless of social class, dressed remarkably alike, in the sartorial equivalent of a monotone. Color in this gray Calvinist land must have struck the eye with unimaginable force—and the color of tulips was like no color anyone had ever laid eyes on before: saturated, brilliant, more intense than that of any other flower.

The story of the Semper Augustus, the most celebrated and expensive tulip for most of the seventeenth century, is a reminder that beauty did in fact underwrite the mania—that, at least in Holland in the 1630s, pork bellies could never have substituted for tulips. The consensus was that Semper Augustus was the most beautiful flower in the world, a masterpiece. "The color is white, with Carmine on a blue base, and with an unbroken flame right to the top," Nicolaes van Wassenaer wrote in 1624 after seeing the tulip in the garden of one Dr. Adriaen Pauw. "Never did a Florist see one more beautiful than this." There were only a dozen or so specimens in existence—and Dr. Pauw owned nearly all of them. This passionate tulip fancier (who was a director of the new East India Company) grew them on his estate in Heemstede, near Haarlem, where he had deployed an elaborate mirrored gazebo in his garden to multiply the effect of his precious blooms. Through the 1620s, Dr. Pauw was bombarded with wildly escalating offers to sell his Semper Augustus bulbs, but he would not part with them at any price. That refusal—which at least one historian credits with igniting the mania—was grounded in the fact that, as Wassenaer tells us, this connoisseur judged the pleasure of looking at a Semper Augustus far superior to any profit.

Before the speculation came the looking.

∷

Looking at my own black tulip, the Queen of Night, here on my desk, I can see it has the classic form of the single tulip: six petals arrayed in two tiers (three inner petals cupped inside three outer ones) that draw an oblong vault of space around the flower's sexual parts, simultaneously advertising and sheltering them from view; each petal is at once a flag and a curtain, drawn. I see too that the petals are not identical: the inner petals have a small, delicate cleft at the top, while the sturdier outer ones form uninterrupted ovals, their incised edges as clean as a blade's. The petals look soft and silky but are not: to the touch they're unexpectedly hard, like orchid petals, and no more silky than this page. Together the six convex petals fit together to form a tailored, somewhat austere blossom; inviting neither touch nor smell, the flower asks me to admire it from a distance. The fact that Queen of Night has no detectable scent is fitting: this is an experience designed strictly for the delectation of the eye.

The long, curving stem of my Queen of Night is nearly as beautiful as the flower it supports. It is graceful, but graceful in a specifically masculine way. This is not the grace of a woman's neck as much as that of a stone sculpture or the curving steel cables of a suspension bridge. The curve seems economical, purposeful, inevitable in its structural logic, even as it changes over time. A horticulturally inclined mathematician would no doubt be able to represent the stem of my tulip in a differential equation.

As the day warms, the curve of the stem relaxes and the petals pull back to reveal the flower's interior space and organs. Like everything else about the tulip, these, too, are explicit and logical. Six stamens—one for every petal—circle around a sturdy upright pedestal, each of them extending, like trembling suitors, a powdery yellow bouquet. Crowning the central pedestal, which botanists call a "style," is the stigma, a pursed set of slightly crooked lips (typically three) poised to receive the grains of pollen, conducting them downward toward the flower's ovary. Sometimes, as now, a single glistening droplet of liquid (nectar? dew?) appears on the stigma's lip, a suggestion of receptiveness.

Everything about tulip sex seems orderly and intelligible; there is none of the occult mystery that attends the sexuality of, say, a Bourbon rose or a doubled peony. Those two are flowers in which one imagines a bumblebee being forced to feel his way around in the dark, stumbling blindly, drunkenly, getting himself all tangled in their innumerable petals. Which is precisely the idea, of course. But it is not the tulip's idea.

In this, I think, lies the key to the distinctive personality of the tulip, if not to the nature of floral beauty in general. Compared to the other canonical flowers, the beauty of the tulip is classical rather than romantic. Or, to borrow the useful dichotomy drawn by the Greeks, the tulip is that rare figure of Apollonian beauty in a horticultural pantheon mainly presided over by Dionysus.

Certainly the rose and peony are Dionysian flowers, deeply sensual and captivating us as much through the senses of touch and smell as sight. The entirely unreasonable multiplication of their petals (one Chinese tree peony is said to have more than three hundred) defies clear seeing and good sense; the profusion of folds edges toward a gorgeous, intoxicating incoherence. To lean in and inhale the breath of a rose or peony is momentarily to leave our rational selves behind, to be transported as only a haunting fragrance can transport us. This is what is meant by ecstasy: to be taken out of ourselves. Such flowers propose a dream of abandon instead of form.

The tulip, by contrast, is all Apollonian clarity and order. It's a linear, left-brained sort of flower, in no way occult, explicit and logical in its formal rules and arrangements (six petals corresponding to six stamens), and conveying all this rationality the only way conceivable: through the eye. The clean, steely stem holds the solitary flower up in the air for our admiration, positing its lucid form over and above the uncertain, shifting earth. The tulip's blooms float above nature's turmoil; even when they expire they do so gracefully. Instead of turning to mush, like a spent rose, or to used Kleenex, like peony petals, the six petals on a tulip cleanly, dryly, and, often simultaneously, shatter.

Friedrich Nietzsche described Apollo, in contrast to Dionysus, as "the god of individuation and just boundaries." Unlike the great mass of flowers, a tulip bloom stands as an individual in the landscape or vase: one bloom per plant, each one perched atop its stem very much like a head. (Recall that the word *tulip* comes from the Turkish word for "turban.") Lower down the figure come the elongated leaves, precisely two in most botanical renderings, often deployed like limbs. It's no surprise that the tulip was the first flower to have its cultivars individually named—and named for individuals.

But unlike most other flowers, which bear female or feminine names, the nomenclature of tulips (Queen of Night notwithstanding) is rife with the names of great men, especially generals and admirals. In the Greek mind the Dionysian was most often associated with the female principle (or at least with androgyny), the Apollonian with the male. Similarly, the Chinese divided flowers, like everything else, into (female) yin and (male) yang. In Chinese thought the soft and extravagantly petaled peony blossom represents the very essence of yin (though its more linear stems and roots are deemed to be yang). Biologically speaking, most flowers (including tulips) are bisexual, containing both male and female organs, yet in our imaginations they tend to lean one way or the other, their forms recalling masculine or feminine beauty and sometimes even male or female organs. There's a rose in my garden, blowsily doubled and colored the palest pink, that the French call Cuisse de Nymph Emue—it was not enough, apparently, to liken this seductive bloom to the "thigh of a nymph," so it became "thigh of an aroused nymph." You can walk through any garden and choose up sides: boy, girl, boy, girl, girl, girl. . . . The canonical flowers seem to me almost all female—except, that is, for the tulip, perhaps the most masculine of flowers. If you doubt this, watch next April how a tulip forces its head up out of the ground, how the head gradually colors as it rises. Dig down along the shaft, and you'll find its bulb, smooth, rounded, hard as a nut, a form for which the botanists offer a most graphic term: "testiculate."

:||:

Of course, like all of our (Apollonian) efforts to order and categorize nature, this one goes only so far before the (Dionysian) pull of things as they really are begins to take its inevitable toll. I mentioned the orderly arrangement of petals and stamens on the Queen of Night on my desk, yet when I went back to the garden to cut another (I have a completely unreasonable number of Queen of Nights in my garden), I noticed for the first time that the bed was teeming with subtle perversities. Here were Queen of Nights with nine and even ten petals, mutant stigmas with six lips instead of three, and in one case a leaf streaked with deep purple, as if its dull green had been infiltrated by the colored petals overhead, their pigment somehow seeping through the plant's body like a dye or drug.

As anyone who grows a lot of them knows, tulips are prone to such eruptions of biological irrationality—chance mutations, color breaks, and instances of "thievery." Thievery is the tulip grower's term for a mysterious phenomenon that causes certain flowers in a field to revert to the form and color of their parent. What I saw in my bed of Nights was an instance of the wondrous instability that inspired the belief that nature liked to play with tulips more than with any other flower.

:||:

A few weeks ago I passed through Grand Army Plaza in Manhattan, where a large flower bed off Fifth Avenue had been planted with thousands of fat yellow Triumphs, arranged with dulling parade-ground precision. They were exactly the sort of stiff, primary-color tulips I used to plant in my parents' yard. I'd read that even today, at a time when tulip growers struggle mightily to keep their fields free of the virus that causes the flower to break, it still occasionally happens. And there in the middle of that relentless, monotonous bed, I spotted one: a violent eruption of red on a chaste canary petal. It wasn't the most handsome of breaks, but the flare of carmine leaping up from the base of that one bloom stood out in the grid of conformists like an exuberant clown, pulling the rug out from under the dream of order this flower bed was meant to represent.

And there was something thrilling about it—I could hardly believe my luck. To me that careless splash of red seemed almost like a visitation—of the distant tulip past, yes, for here was the return of the virus so assiduously repressed, but of something else, too, some inchoate, underground force that riveted me. It was as if the whole grid of flowers and, by extension, the grid of the city itself had been put in doubt by that one ecstatic, wayward pulse of life. (Or was it death? I guess you'd have to say it was both.)

Then, that night, I dreamt about what I'd witnessed, the stiff yellow grid and its solitary red joker. In the dream version the broken tulip appears in the front row, and right beside it lies a fancy fountain pen, a Montblanc. (This is all too embarrassing to make up.) In a gesture of impetuousness completely out of character, I grab them both, the broken tulip and the pen, and run like a man possessed up Fifth Avenue. I'm flying by the spinning doors of the Plaza and Pierre hotels when I snag the attention of the two brass-buttoned doormen standing sentry outside the Pierre. They can have no idea who I am or what I've done, but they leap to and give slapstick chase anyway, their cartoon hollerings—"Stop! Thief!"—sounding in my ears as I tear up the avenue, clutching my tulip and pen and laughing hysterically at the absurdity of it all—the circumstance, but also the dream about it.

∷

Color breaks far more beautiful than the one I saw on Fifth Avenue had helped fire the tulipomania, a speculative frenzy that, like the breaks themselves, can perhaps best be understood as an explosive outbreak of the Dionysian in the too-strict Apollonian world of the tulip—and of the Dutch bourgeoisie. This, at least, is how I've come to think of the tulipomania—as a festival of Dionysus, by turns ecstatic and destructive, transplanted from the forest or temple to the orderly precincts of the marketplace.

Tulipomania bore all the hallmarks of a medieval carnival, in which, for a brief "orgasmic interim" (in the words of the French historian Le Roy Ladurie), the stable order of society was turned on its head. A carnival is a social ritual of sanctioned craziness and release—a way for a community to temporarily indulge its Dionysian urges. For its duration, the identity of everyone swept into its vortex is up for grabs: the village idiot is made king, the poor man suddenly rich, the rich man just as suddenly a pauper. Everyday roles and values are suddenly, thrillingly, suspended, and astounding new possibilities arise.

As with society, so with capitalism in the throes of a speculative mania: all of its values are turned on their head—thrift, patience, value for money, reward for effort. For as long as the carnival of capitalism lasts, the rules of logic are repealed, or rather recast along new lines, ones that will appear absurd in the cold light of the morning after but that make impeccable sense within the fevered space of the speculative bubble.

It's hard to date with precision exactly when the bubble in Holland formed, but the autumn of 1635 marked a turning point. That's when the trade in actual bulbs gave way to the trade in promissory notes: slips of paper listing details of the flowers in question, the dates they would be delivered, and their price. Before then, the tulip market followed the rhythm of the season: bulbs could change hands only between the months of June, when they were lifted from the ground, and October, when they had to be planted again. Frenzied as it was, the market before 1635 was still rooted in reality: cash money for actual flowers. Now began the *windhandel*—the wind trade.

Suddenly the tulip trade was a year-round affair, and the connoisseurs and growers who shared a genuine interest in the flowers were joined by legions of newly minted "florists" who couldn't have cared less. These men were speculators who, only days before, had been carpenters and weavers, woodcutters and glass-blowers, smiths, cobblers, coffee grinders, farmers, tradesmen, peddlers, clergymen, schoolmasters, lawyers, and apothecaries. One burglar in Amsterdam pawned the tools of his trade so that he too could become a speculator in tulips.

Rushing to get in on the sure thing, these people sold their businesses, mortgaged their homes, and invested their life savings in slips of paper representing future flowers. Predictably, the flood of fresh capital into the market drove prices to bracing new heights. In the space of a month the price of a red-and-yellow-striped Gheel ende Root van Leyden leapt from 46 guilders to 515. A bulb of Switsers, a yellow tulip feathered with red, soared from 60 to 1,800 guilders.

At its height, the trade in tulips was conducted by florists in "colleges"—back rooms of taverns given over to the new business two or three days a week. Colleges quickly developed a set of rituals that sound like a cross between orderly stock market protocol and a drinking contest. Under one common set of procedures, called *met de borden*, or "with the boards," a seller and buyer who wanted to do business were handed slates on which they wrote an opening price for the tulip in question. The slates were then passed to a pair of proxies (essentially arbitrators nominated by the traders), who would then settle on a price somewhere between the two opening bids; this they would scribble on the slates before passing them back to the principals. The traders could either let the number stand, signifying agreement, or rub it out. If both rubbed out the price, the deal was off; but if only one party declined, that florist had to pay a fine to the college—an incentive to close the deal. When a deal did close, the buyer had to pay a small commission, called the *wijnkoopsgeld*: wine money. In keeping with the carnival atmosphere, these fines and commissions were used to buy wine and beer for everyone—another incentive to make deals. In a satirical pamphlet describing the scene, an old-timer advises his neophyte friend to drink up: "This trade must be done with an intoxicated head, and the bolder one is, the better."

∷

The bubble logic driving tulipomania has since acquired a name: "the greater fool theory." Although by any conventional measure it is folly to pay thousands for a tulip bulb (or for that matter an Internet stock), as long as there is an even greater fool out there willing to pay even more, doing so is the most logical thing in the world. By 1636 the taverns were crowded with such people, and as long as Holland remained home to an expanding population of greater fools—people blinded by their desire for instant wealth—the truly foolish act would have been to abstain from the tulip trade.*

---

*It might also be that, for some of the Calvinist Dutch, financial abandon offered a way to atone for what they felt was the shame of their wealth, the embarrassment of their riches: they were trading their filthy lucre for the pristine beauty of a flower.

Even so, there was more to the *windhandel* than mere wind. For the tulip craze marked the birth of a real business—the Dutch bulb trade—that would long outlast the mania. (The same could be said of our own Internet bubble: beneath the froth of speculation is a new and important industry.) According to Joseph Schumpeter, it is not at all unusual for the birth of a new business to be attended by a speculative bubble as capital rushes in, dazzled by the young industry's wildly exaggerated promise.

Every bubble sooner or later must burst—the carnival that was permanent would spell the end of the social order. In Holland the crash came in the winter of 1637, for reasons that remain elusive. But with real tulips about to come out of the ground, paper trades and futures contracts would soon have to be settled—real money would soon have to be exchanged for real bulbs—and the market grew jittery.

On February 2, 1637, the florists of Haarlem gathered as usual to auction bulbs in one of the tavern colleges. A florist sought to begin the bidding at 1,250 guilders for a quantity of tulips—Switsers, in one account. Finding no takers, he tried again at 1,100, then 1,000 . . . and all at once every man in the room—men who days before had themselves paid comparable sums for comparable tulips—understood that the weather had changed. Haarlem was the capital of the bulb trade, and the news that there were no buyers to be found there ricocheted across the country. Within days tulip bulbs were unsellable at any price. In all of Holland a greater fool was no longer to be found.

In the aftermath, many Dutch blamed the flower for their folly, as if the tulips themselves had, like the sirens, lured otherwise sensible men to their ruin. Broadsides excoriating the tulipomania became best-sellers: *The Fall of the Great Garden-Whore, the Villain-Goddess Flora; Flora's Fool's Cap, or Scenes from the Re-*

markable Year 1637 when one Fool hatched another, the Idle Rich lost their wealth and the Wise lost their senses; Charge Against the Pagan and Turkish Tulip-Bulbs. (Flora was, of course, the Roman goddess of flowers, who was a prostitute famous for bankrupting her lovers.) In the months after the fever broke, a professor of botany at the University of Leiden, a man named Fortius who occupied Clusius's old chair, could be seen patrolling the streets of the city, beating any tulip he encountered with his cane. At the conclusion of a medieval carnival, it was the carnival king who was hung in effigy. Likewise, the ancient festivals of Dionysus would end in destruction and mutilation and the sacrifice of the god himself.

⠶

It bears remembering that tulipomania was finally a frenzy not of consumption or of pleasure but of financial speculation, and that it took place not in a country ordinarily given to large passions but rather in the most stolid bourgeois culture of the time. The Dionysian eruptions of the tulip are relative, in other words, making an impression in direct proportion to their anomalousness.

Certainly the color break I spotted in Grand Army Plaza was like that—a wayward splatter of paint on a monochromatic ground, an extravagance I might not have noticed if not for the scrupulous precinct of order—of petal, of blossom, of plant—in which it happened to detonate. Etymologically, the word *extravagant* means to wander off a path or cross a line—orderly lines, of course, being Apollo's special domain. In this may lie a clue to the abiding power of the tulip, as well as, perhaps, to the nature of beauty. The tulip is a flower that draws some of the most exquisite lines in nature and then, in spasms of extravagance, blithely oversteps them. On the same principle, syncopation enlivens a regular, four-four measure of music, enjambment the stately line of iambic pentameter. So here is a third constituent of beauty to add to the desiderata offered to us by the flower: first came contrast, then pattern (or form), and finally variation.

The pleasure we take in the breaking of a too-predictable pattern may account for the allure of the broken tulip, as well as the Rembrandt and the parrot (a type of tulip that explodes the tailored flower into the exuberant frills of a party dress). Then, of course, there is the black tulip, the gothic femme fatale in the masculine world of tulips. In the Queen of Night the mysteriously depthless hue plays against the sunny lucidity of her form. Our eyes and ears quickly tire of any strict Apollonian order that isn't shadowed by some hint, some threat, of trespass or waywardness.

By the same token, the most breathtaking rose or peony is the one in which the tumbling profusion of its petals is held in check by some kind of form or frame; the slightest suggestion of symmetry—the form of a globe or teacup, say—keeps the bloom from going slack. The Greeks believed that true beauty (as opposed to mere prettiness) was the offspring of these two opposing tendencies, which they personified in Apollo and Dionysus, their two gods of art. Great art is born when Apollonian form and Dionysian ecstasy are held in balance, when our dreams of order and abandon come together. One tendency uninformed by the other can bring forth only coldness or chaos—the stiffness of a Triumph tulip, the slackness of a wild rose. So though we can classify any particular flower as Apollonian or Dionysian (or male or female) the most beautiful flowers—like Semper Augustus or Queen of Night—are the ones that also partake of their opposing element.

The Greeks' myth of beauty, the most persuasive I know of, takes us most of, but not all, the way back to beauty's origins in the commingling of tendencies found in the human brain and breast. But the birth of beauty goes back further still, to a time before Apollo and Dionysus, before human desire, when the world was mostly leaf and the first flower opened.

⫶

Once upon a time, there were no flowers—two hundred million years ago, to be only slightly more precise. There were plants then, of course, ferns and mosses, conifers and cycads, but these plants didn't form true flowers or fruit. Some of them reproduced asexually, cloning themselves by various means. Sexual reproduction was a relatively discreet affair usually accomplished by releasing pollen onto the wind or water; by sheer chance some of it would find its way to other members of the species, and a tiny, primitive seed would result. This prefloriferous world was a slower, simpler, sleepier world than our own. Evolution proceeded more slowly, there being so much less sex, and what sex there was took place among close-by and closely related plants. Such a conservative approach to reproduction made for a biologically simpler world, since it generated relatively little novelty or variation. Life on the whole was more local and inbred.

The world before flowers was sleepier than ours because, lacking fruit and large seeds, it couldn't support many warm-blooded creatures. Reptiles ruled, and life slowed to a crawl whenever it got cold; little happened at night. It was a plainer-looking world, too, greener even than it is now, absent all the colors and patterns (not to mention scents) that flowers and fruits would bring into it. Beauty did not yet exist. That is, the way things looked had nothing to do with desire.

Flowers changed everything. The angiosperms, as botanists call the plants that form flowers and then encased seeds, appeared during the Cretaceous period, and they spread over the earth with stunning rapidity. "An abominable mystery" is how Charles Darwin described this sudden and entirely evitable event. Now, instead of relying on wind or water to move genes around, a plant could enlist the help of an animal by striking a grand co-evolutionary compact: nutrition in exchange for transportation. With the advent of the flower, whole new levels of complexity come into the world: more interdependence, more information, more communication, more experimentation.

The evolution of plants proceeded according to a new motive force: attraction between different species. Now natural selection favored blooms that could rivet the attention of pollinators, fruits that appealed to foragers. The desires of other creatures became paramount in the evolution of plants, for the simple reason that the plants that succeeded at gratifying those desires wound up with more offspring. Beauty had emerged as a survival strategy.

The new rules speeded the rate of evolutionary change. Bigger, brighter, sweeter, more fragrant: all these qualities were quickly rewarded under the new regime. But so was specialization. Since bestowing one's pollen on an insect that might deliver it to the wrong address (such as the blossoms of unrelated species) was wasteful, it became an advantage to look and smell as distinctive as possible, the better to command the undivided attention of a single, dedicated pollinator. Animal desire was thus parsed and subdivided, plants specialized accordingly, and an extraordinary flowering of diversity took place, much of it under the signs of co-evolution and beauty.

With flowers came fruit and seeds, and these, too, remade life on Earth. By producing sugars and proteins to entice animals to disperse their seed, the angiosperms multiplied the world's supply of food energy, making possible the rise of large warm-blooded mammals. Without flowers, the reptiles, which had gotten along fine in a leafy, fruitless world, would probably still rule. Without flowers, we would not be.

∷

So the flowers begot us, their greatest admirers. In time human desire entered into the natural history of the flower, and the flower did what it has always done: made itself still more beautiful in the eyes of this animal, folding into its very being even the most improbable of our notions and tropes. Now came roses that resembled aroused nymphs, tulip petals in the shape of daggers, peonies bearing the scent of women. We in turn did our part, multiplying the flowers beyond reason, moving their seeds around the planet, writing books to spread their fame and ensure their happiness. For the flower it was the same old story, another grand coevolutionary bargain with a willing, slightly credulous animal—a good deal on the whole, though not nearly as good as the earlier bargain with the bees.

And what about us? How did we make out? We did very well by the flower. There were, of course, the pleasures to the senses, the sustenance of their fruit and seeds, and the vast store of new metaphor. But we gazed even farther into the blossom of a flower and found something more: the crucible of beauty, if not art, and maybe even a glimpse into the meaning of life. For look into a flower, and what do you see? Into the very heart of nature's double nature—that is, the contending energies of creation and dissolution, the spiring toward complex form and the tidal pull away from it. Apollo and Dionysus were names the Greeks gave to these two faces of nature, and nowhere in nature is their contest as plain or as poignant as it is in the beauty of a flower and its rapid passing. There, the achievement of order against all odds and its blithe abandonment. There, the perfection of art and the blind flux of nature. There, somehow, both transcendence *and* necessity. Could that be it—right there, in a flower—the meaning of life?

CHAPTER 3

Desire: Intoxication

Plant: Marijuana

(CANNABIS SATIVA × INDICA)

The forbidden plant and its temptations are older than Eden, go back further even than we do. So too the promise, or threat, that forbidden plants have always made to the creature who would taste them—the promise, that is, of knowledge and the threat of mortality. If it sounds as if I'm speaking metaphorically about forbidden plants and knowledge, I don't mean to. In fact, I'm no longer so sure the author of Genesis was, either.

Living things have always had to make their way in a wild garden of flowers and vines, of leaves and trees and fungi that hold out not only nourishing things to eat but deadly poisons, too. Nothing is more important to a creature's survival than knowing which is which, yet drawing a bright line through the middle of the garden, as the God of Genesis found, doesn't always work. The difficulty is that there are plants that do other, more curious things than simply sustain or extinguish life. Some heal; others rouse or calm or quiet the body's pain. But most remarkable of all, there are plants in the garden that manufacture molecules with the power to change the subjective experience of reality we call consciousness.

Why in the world should this be so—why should evolution yield plants possessing such magic? What makes these plants so irresistible to us (and to many other creatures), when the cost of using them can be so high? Just what is the knowledge held out by a plant such as cannabis—and why is it forbidden?

:‖:

Start with the bright line, as all creatures must. How does one tell the dangerous plants from the ones that merely nourish? Taste is the first tip-off. Plants that don't wish to be eaten often manufacture bitter-tasting alkaloids; by the same token, plants that do wish to be eaten—like the apple—often manufacture a super-abundance of sugars in the flesh around their seeds. So as a general rule, sweet is good, bitter bad. Yet it turns out that it is some of the bitter, bad plants that contain the most powerful magic—that can answer our desire to alter the textures and even the contents of our consciousness. There it is, right in the middle of the word *intoxication*, hidden in plain sight: *toxic*. The bright line between food and poison might hold, but not the one between poison and desire.

:‖:

The manifold and subtle dangers of the garden, to which a creature's sense of taste offers only the crudest map, are mainly the fruits of strategies plants have devised to defend themselves from animals. Most of the ingenuity of plants—that is, most of the work of a billion years of evolutionary trial and error—has been applied to learning (or rather, inventing) the arts of biochemistry, at which plants excel beyond all human imagining. (Even now a large part of human knowledge about making medicines comes directly from plants.) While we animals were busy nailing down things like locomotion and consciousness, the plants, without ever lifting a finger or giving it a thought, acquired an array of extraordinary and occasionally diabolical powers by discovering how to synthesize remarkably complicated molecules. The most remarkable of these molecules (at least from our perspective) are the ones designed expressly to act on the brains of animals, sometimes to attract their attention (as in the scent of a flower) but more often to repel and sometimes even destroy them.

Some of these molecules are outright poisons, designed simply to kill. But one of the great lessons of coevolution (a lesson recently learned by designers of pesticides and antibiotics) is that the all-out victory of one species over another is often Pyrrhic. That's because a powerful, death-dealing toxin can exert such a strong selective pressure for resistance in its target population that it is quickly rendered ineffective; a better strategy may be to repel, disable, or confound. This fact might explain the astounding inventiveness of plant poisons, the vast catalog of chemical curiosities and horrors that first flowered in Cretaceous times with the rise of the angiosperms. The same evolutionary watershed—Darwin's "abominable mystery"—that ushered in the dazzling arts of floral attraction brought with it the darker arts of chemical warfare.

Some plant toxins, such as nicotine, paralyze or convulse the muscles of pests who ingest them. Others, such as caffeine, unhinge an insect's nervous system and kill its appetite. Toxins in datura (and henbane and a great many other hallucinogens) drive a plant's predators mad, stuffing their brains with visions distracting or horrible enough to take the creatures' mind off lunch. Compounds called flavonoids change the taste of plant flesh on the tongues of certain animals, rendering the sweetest fruit sour or the sourest flesh sweet, depending on the plant's designs. Photosensitizers present in species such as the wild parsnip cause the animals that eat it to burn in the sun; chromosomes exposed to these compounds spontaneously mutate when exposed to ultraviolet light. A molecule present in the sap of a certain tree prevents caterpillars that sample its leaves from ever growing into butterflies.

By trial and error animals figure out—sometimes over eons, sometimes over a single lifetime—which plants are safe to eat and which forbidden. Evolutionary counterstrategies arise too: digestive processes that detoxify, feeding strategies that minimize the dangers (like that of the goat, which nibbles harmless quantities of a great many different plants), or heightened powers of observation and memory. This last strategy, at which humans particularly excel, allows one creature to learn from the mistakes and successes of another.

The "mistakes" are, of course, especially instructive, as long as they're not your own or, if they are, they prove less than fatal. For even some of the toxins that kill in large doses turn out in smaller increments to do interesting things—things that are interesting to animals as well as people. According to Ronald K. Siegel, a pharmacologist who has studied intoxication in animals, it is common for animals deliberately to experiment with plant toxins; when an intoxicant is found, the animal will return to the source repeatedly, sometimes with disastrous consequences. Cattle will develop a taste for locoweed that can prove fatal; bighorn sheep will grind their teeth to useless nubs scraping a hallucinogenic lichen off ledge rock. Siegel suggests that some of these adventurous animals served as our Virgils in the garden of psychoactive plants. Goats, who will try a little bit of anything, probably deserve credit for the discovery of coffee: Abyssinian herders in the tenth century observed that their animals would become particularly frisky after nibbling the shrub's bright red berries. Pigeons spacing out on cannabis seeds (a favorite food of many birds) may have tipped off the ancient Chinese (or Aryans or Scythians) to that plant's special properties. Peruvian legend has it that the puma discovered quinine: Indians observed that sick cats were often restored to health after eating the bark of the cinchona tree. Tukano Indians in the Amazon noticed that jaguars, not ordinarily herbivorous, would eat the bark of the yaje vine and hallucinate; the Indians who followed their lead say the yaje vine gives them "jaguar eyes."

:||:

Whenever I read something like this, I wonder, How do you tell when a jaguar is hallucinating? Then I think about Frank, my late, cranky old tomcat, who I became convinced used drug plants habitually in order to hallucinate. Every summer evening at around five, Frank would lumber into the vegetable garden for a happy-hour nip of *Nepeta cataria*, or catnip. He would first sniff, then tug at the leaves with his teeth and proceed to roll around in paroxysms of what looked to me like sexual ecstasy. His pupils would shrink to pinpricks and take on a slightly scary thousand-mile stare, preparatory to pouncing on unseen enemies or—who can say?—lovers. Frank would crash-land in the dirt, pick himself up, do a funny little sidestep, then pounce again until, exhausted, he'd go sleep it off in the shade of a tomato plant.

I learned later that catnip contains a chemical compound, called "nepetalactone," which mimics the pheromone cats produce in their urine during courtship. This chemical key just happens to fit an aphrodisiac lock in a cat's brain and apparently no other. It was amusing to watch a plant derange my cat, but also unsettling; for that brief interlude, Frank would wobble through the garden as though he were literally beside himself. Yet he'd be back again the next day—though, curiously, never before five. Maybe he ritualized the practice to keep it under control; or maybe it took him the better part of the day to remember just where it was that the magic plant grew.

I'd planted the catnip strictly for Frank's pleasure, though looking back I sometimes wonder if the plant wasn't also in my garden as a substitute, or placeholder, for the forbidden plant I sometimes wished I could grow for myself. Cannabis, I mean. At once an intoxicant, a medicine, and a fiber (this last use, admittedly, of absolutely no interest to me), cannabis is one of the most powerful of the plants that will grow around here; it is also, as I write, the most dangerous plant I could grow in my garden. Frank's happy-hour ritual was a daily reminder that my garden was capable of producing much more than food or beauty, that it also could perform some rather remarkable feats of brain chemistry and by doing so answer other, more complicated desires.

:|:

I sometimes think we've allowed our gardens to be bowdlerized, that the full range of their powers and possibilities has been sacrificed to a cult of plant prettiness that obscures more dubious truths about nature, our own included. It hasn't always been this way, and we may someday come to regard the contemporary garden of vegetables and flowers as a place almost Victorian in its repressions and elisions.

For most of their history, after all, gardens have been more concerned with the power of plants than with their beauty—with the power, that is, to change us in various ways, for good and for ill. In ancient times, people all over the world grew or gathered sacred plants (and fungi) with the power to inspire visions or conduct them on journeys to other worlds; some of these people, who are sometimes called shamans, returned with the kind of spiritual knowledge that underwrites whole religions. The medieval apothecary garden cared little for aesthetics, focusing instead on species that healed and intoxicated and occasionally poisoned. Witches and sorcerers cultivated plants with the power to "cast spells"—in our vocabulary, "psychoactive" plants. Their potion recipes called for such things as datura, opium poppies, belladonna, hashish, fly-agaric mushrooms (*Amanita muscaria*), and the skins of toads (which can contain DMT, a powerful hallucinogen). These ingredients would be combined in a hempseed-oil-based "flying ointment" that the witches would then administer vaginally using a special dildo. This was the "broomstick" by which these women were said to travel.

The medieval gardens of witches and alchemists were forcibly uprooted and forgotten (or at least euphemized beyond recognition), but even the comparatively benign ornamental gardens that came after them went out of their way to honor the darker, more mysterious face of nature. The Gothic gardens of England and Italy, for example, always made room for intimations of mortality—by including a dead tree, say, or a melancholy grotto—and the occasional frisson of horror. These gardens were interested in changing people's consciousness, too, though more in the way a horror movie does than a drug. It's only been in modern times, after industrial civilization concluded (somewhat prematurely) that nature's powers were no longer any match for its own, that our gardens became benign, sunny, and environmentally correct places from which the old horticultural dangers—and temptations—were expelled.

Or if not expelled, almost willfully forgotten. For even in

Grandmother's garden you're apt to find datura and morning glories (the seeds of which some Indians consume as a sacramental hallucinogen) and opium poppies—right there, the makings of a witch's flying ointment or apothecary's tonic. The knowledge that once attended these powerful plants, however, has all but vanished. And as soon as this plant knowledge is restored to consciousness—as soon as, say, one forms the intention of slitting the head of an opium poppy to release its narcotic sap—so too must be its taboo. Curiously, growing *Papaver somniferum* in America is legal—unless, that is, it is done in the knowledge that you are growing a drug, when, rather magically, the exact same physical act becomes the felony of "manufacturing a controlled substance." Evidently the Old Testament and the criminal code both make a connection between forbidden plants and knowledge.

:||:

I once grew opium poppies in my garden—yes, with felonious intent. I also grew marijuana, back when that was no big deal. I still grow grapes and hops, both of which can be made into legal intoxicants (as long as I don't sell them), and, in my herb garden, Saint-John's-wort (an antidepressant), chamomile, and valerian (both mild sedatives).

I should probably explain my interest in these plants. At least in the beginning, this had less to do with my interest in using drugs, which was never more than mild, than with an impulse I think most gardeners share. In fact, by the time I planted a few cannabis seeds, in the early 1980s, I no longer smoked at all—pot, fairly reliably, rendered me paranoid and stupid. But I had just taken up gardening and was avid to try anything—the magic of a Bourbon rose or a beefsteak tomato seemed very much of a piece with the magic of a psychoactive plant. (I still feel this way.) So when my sister's boyfriend asked if I might want to plant a few seeds he'd picked out of "some really amazing Maui," I decided to give it a try—as much as anything, just to see if I could grow it.

To another gardener, this will not seem odd, for we gardeners are like that: eager to try the improbable (if only to harvest a good story), to see if we can't grow an artichoke in zone five or brew homemade echinacea tea from the roots of our purple coneflowers. Deep down I suspect that many gardeners regard themselves as small-time alchemists, transforming the dross of compost (and water and sunlight) into substances of rare value and beauty and power. Maybe at some level we're still in touch with the power of the old gardens. Also, one of the attractions of gardening is the independence it can confer—from the greengrocer, the florist, the pharmacist, and, for some, the drug dealer. One does not have to go all the way "back to the land" to experience the satisfaction of providing for yourself off the grid of the national economy. So, yes, I was curious to see if I could grow some "really amazing Maui" in my Connecticut garden. It seemed to me this would indeed represent a particularly impressive sort of alchemy. But as things turned out, my experiment in growing marijuana was of a piece with my experience smoking it, paranoid and stupid being the operative terms.

⫶

It was in the spring of 1982, I believe, that I sprouted a handful of the Maui seeds on a moistened paper towel; within days two of them had germinated. As soon as the weather warmed, I planted the seedlings outdoors, not in the garden proper but behind the falling-down barn back behind the house, in a mound of ancient cow manure I had inherited from the dairy farmer whose place this used to be.

I more or less forgot about the plants until a few months

later, when I returned to find what looked like a pair of Christmas trees, eight feet tall at least, rising over the late-summer weeds—lush, leafy, emerald green shrubs growing avidly in the thinning September light. No one would ever claim marijuana is a great beauty, though a gardener can't help but admire the sheer green exuberance of this plant, a towering heap of leafy palms held up to the sun in an ecstatic frenzy of photosynthesis. The plant has the ardor of a weed.

Though frost was just around the corner (I've lost tomatoes here as early as September 15), the big plants gave no sign that they were even thinking of flowering. This I regarded as disappointing but hardly tragic, since in those days people still smoked cannabis leaves. (Nowadays, of course, only the unpollinated female flowers—called sinsemilla—are deemed worthwhile; growers simply throw the leaves and stems onto the compost pile.) Even so, I decided to hold off for a few more weeks to see if I couldn't harvest a few buds.

The plants continued to grow at an alarming rate, adding as much as a foot to their height and girth every week, so that by the end of September they'd made themselves conspicuous from just about any point on the property. There they were, a couple of jolly green giants lurking behind the barn—and I found myself in a state of almost perpetual anxiety and dread. I'd read in the papers that the state police sometimes did aerial reconnaissance to locate marijuana gardens, and anytime I heard the drone of a small plane overhead, I raced outside to see if its flight path would take it over my plants. The slowing down of any full-size American sedan on my road was enough to rattle me. Every day that fall I weighed the risks of detection, and a killing frost, against the potential reward of a few buds.

A close call ended my career as a marijuana farmer. I'd ordered a cord of wood from a man who'd posted a flyer in town. He showed up with the first half early on a Saturday morning, a compact block of a man with a pewter crew cut, and asked where I wanted it stacked. Though open to the elements on two sides, the ruined barn did at least have a tight roof, and we agreed it was far and away the best place to stack the wood. But before getting down to work, the man and I fell into conversation, leaning there on the warm hood of his truck, enjoying the crisp October morning. Making small talk, I asked if he sold cordwood for a living. No, he chuckled, firewood was just a sideline, that and plowing driveways in the winter.

"Nine to five, I'm chief of police of New Milford."

All at once the bones in my legs began to go soft. I found I could no longer form a sentence without specifically addressing the muscles in my lips. The barn, you see, was nothing more than a shell of boards, and no police officer standing in it could fail to spot the two green giants through the opening in the rear wall. But what else was I going to do? Dumping the wood anywhere *but* in the barn was ridiculous.

Unfortunately, no nonridiculous stratagem presented itself to my stupefied brain. I simply blurted out that, on second thought, I wanted the whole load dumped right here in the middle of the driveway, that'd be just fine, thanks.

"Don't be silly," the police chief said, turning to climb back into the truck's cab. "It's no trouble at all. I'll just back the load up to the barn."

"Uh . . . no!" I can only imagine how I must have sounded. "Right here, here is perfect. Near the house . . . burn it right away."

"Okay, maybe *some* of it, but not the whole cord." The truck's engine roared to life.

"Yes! The whole cord! *Here!*" I may have been shouting now. "This is exactly where I want it!" And before he could throw his transmission into reverse, I jumped up onto the rear fender and started furiously to throw logs over my shoulder, onto the driveway and the lawn behind the truck, anywhere to block its path to the barn. The man got out, squinted at me in bewilderment, and then, finally, blessedly, shrugged his shoulders. The words "Suit yourself" have never sounded so sweet.

As soon as the wood was unloaded, the chief of police drove off to go get the second half of my cord, and I, temporarily reprieved but still in full panic, ransacked the toolshed in search of an ax. There would be no buds after all. I chopped down the two plants, the trunks of which were as thick around as my forearms, hacked off the branches, and stuffed the fragrant mass of foliage into a pair of heavy-duty trash bags, which I hauled up to a crawl space in the attic—all in about four minutes. My harvest, when dried, yielded a couple of pounds of leaves that smelled like old socks. *Something* happened when you smoked them, but the effect had less in common with a high than with a sinus headache.

⋮⋮

As you can probably guess, I've told my marijuana-growing story more than a few times, after dinner with friends, say, and I can usually count on a few laughs. The happy ending is one reason, but the other reason the story qualifies as light comedy is that the suspense on which it hinges, while real enough, is not exactly a matter of life or death. If the police chief had spotted my plants, things would have gotten uncomfortable for me, but it was not as if I would have gone to jail. In 1982 a legal slap on the wrist, and perhaps a certain amount of personal embarrassment (What do I tell my parents? My boss?) was really about all a small-time marijuana grower had to fear. It hadn't been many years before this misadventure, after all, that an American president—Jimmy Carter—had proposed that marijuana be decriminalized (his sons and even his *drug czar* smoked), and Bob Hope was telling benign jokes about doobies in prime time. Marijuana then was harmless, funny, and, it seemed to everyone, on the verge of social acceptance.

In the years since, there has been a sea change concerning cannabis in America. By the end of the decade the plant had suddenly acquired, or been endowed with, extraordinary new powers, which, among other things, rendered my story a period piece, quaint in its goofiness and not at all likely to be repeated. A couple of facts will illustrate the change: The minimum penalty for the cultivation of a kilogram of marijuana (the size of my harvest, more or less) in this state has, since 1988, been a mandatory five-year jail sentence. (Other states are harsher still: growing *any* amount of marijuana in Oklahoma qualifies a gardener for a life sentence.)

Jail time would not be my only worry were I so foolish as to reprise my experiment. If the New Milford police chief happened to find marijuana growing in my garden today, he would have the power to seize my house and land, regardless of whether I was ultimately convicted of a crime. That's because, according to the somewhat magical reasoning of the federal asset-forfeiture laws, my *garden* can be found guilty of violating the drug laws even if I am not. The titles of proceedings brought under these laws sound rather less like exercises in American jurisprudence than medieval animism: *United States v. One 1974 Cadillac Eldorado Sedan*. If the police chief chose to bring such an action (*The People of Connecticut v. Michael Pollan's Garden*), he'd simply have to prove that my land had been used in the commission of a crime for it to become the property of the New Milford Police Department, theirs to dispose of as they wished. So do things stand in America today that yielding to the temptation of a forbidden plant not only can get you temporarily expelled from your garden but can get your garden taken away forever.

The swiftness of this change in the weather, the demonizing of a plant that less than twenty years ago was on the cusp of general acceptance, will surely puzzle historians of the future. They will wonder why it was that the "drug war" of the '80s, '90s, and '00s fought the vast majority of its battles over marijuana.* They will wonder why, during this period, Americans jailed more of their citizens than any other country in history, and why one of every three of those were in prison because of their involvement with drugs, nearly fifty thousand of them solely for crimes involving marijuana. And they will wonder why Americans would have been willing to give up so many of their hard-won liberties in the fight against this plant. For in the last years of the twentieth century a series of Supreme Court cases and government actions specifically involving marijuana led to a substantial increase in the power of the government at the expense of the Bill of Rights.† As a result of the war against cannabis, Americans are demonstrably less free today.

---

*More drug arrests are for crimes involving marijuana than any other drug: nearly 700,000 in 1998, 88 percent of them for possession. Marijuana cases account for most of the asset forfeitures that law enforcement budgets have come to rely on. Marijuana is the primary focus of drug prevention efforts in the schools, drug testing in the workplace, and public service advertising about drugs.

†What a dissenting Supreme Court justice in 1988 deplored as a new "drug exception to the Constitution" has been substantially based on marijuana cases. For example, in *Illinois v. Gates* (1983) the Supreme Court carved broad new exceptions to the Fourth Amendment right against unreasonable searches, as well as the Sixth Amendment right to confront one's accusers. The venerable principle of

Historians of the future will decide for themselves exactly why marijuana, of all drugs, should have become the focus of the American drug war—why the bright line of prohibition was drawn around this particular plant, rather than coca or poppies. Did marijuana pose a grave threat to public health, or was marijuana the only illicit drug in wide enough use to justify waging so ambitious a war in the first place?* Whatever the case, it's hard to believe such a powerful new taboo against marijuana would have stuck if the plant hadn't already been a powerful symbol. Certainly marijuana's close identification with the counterculture made it an attractive target to a drug war that, whatever else it may have been, was part of a political and cultural reaction against the sixties. But whatever the reason, by the end of the twentieth century this plant and its taboo had appreciably changed American life not

---

posse comitatus, which holds that the armed forces of the United States cannot be used to police U.S. territory, has been suspended during the war against marijuana, notably by President Reagan, who deployed troops to rout out growers in northern California. The First Amendment has suffered as well: magazines aimed at pot growers have been harassed and, in one case (*Sinsemilla Tips*), raided and closed down. In 1998 the federal government threatened to revoke the license of California doctors who exercised their First Amendment right to talk to patients about the medical benefits of marijuana. Also that year, Congress ordered the District of Columbia not to count the votes of its citizens in a referendum on medical marijuana. Arguably, the war against cannabis has also eroded the Sixth Amendment right to a jury trial (since drastic mandatory minimum sentences force most marijuana defendants to accept plea bargains) as well as the presumption of innocence (since asset forfeiture allows the government to seize assets without proving guilt).

*Remove the twenty million or so Americans who use marijuana, and we are left with a "drug abuse epidemic" involving roughly two million regular heroin and cocaine users—a public health problem, to be sure, but serious enough to justify spending $20 billion a year (or modifying the Bill of Rights)?

once but twice: the first time rather mildly, with marijuana's widespread popularity beginning in the sixties, and then again, perhaps more profoundly, in its role as *casus belli* in the war against drugs.

∷

There has been another dramatic change in the story of marijuana since my brief career as a grower, and that is the change in the plant itself. When the natural history of cannabis is written, the American drug war will loom as one of its most important chapters, on a par with the introduction of cannabis to the Americas by African slaves, say, or the ancient Scythians' discovery that hemp could be smoked.* For the modern prohibition against marijuana led directly to a revolution in both the genetics and the culture of the plant. It stands as one of the richer ironies of the drug war that the creation of a powerful new taboo against marijuana led directly to the creation of a powerful new plant.

Marijuana's recent natural history is much harder to reconstruct than its social history, since so much of it took place underground and in secret; this plant's Johnny Appleseeds have tended to be far-flung and anonymous. But I was inspired to go looking for them a few years ago, after I learned (from a friend of a friend) just how sophisticated marijuana cultivation had become in the years since my feeble attempt and how much more potent American pot had grown. This fellow had once helped design and

---

*The practice of smoking as we know it wouldn't come to Europe until Columbus brought it back from America, but the Scythians invented something like it around 700 B.C. According to Herodotus, they would put their heads into small tents designed to trap the fumes from cannabis buds placed on red-hot rocks—"until they rise up to dance and betake themselves to singing."

install a series of state-of-the-art "grow rooms." As I listened to him talk about his work one evening, dilating on the relative benefits of sodium and metal halide lights, the optimal number of clones to plant per kilowatt, and the intricacies of hybridizing *indica*s and *sativa*s, it dawned on me that *this* was what the best gardeners of my generation had been doing all these years: they had been underground, perfecting cannabis.

∷

To a marijuana grower, Amsterdam in the 1990s was something like what Paris in the 1920s was to a writer: a place where alienated expatriates could go to practice their craft in peace and hook up with a community of kindred souls. Growing marijuana is not precisely legal in Holland, but several hundred "coffee shops" are licensed to sell it, and small-scale growing to supply those shops is officially tolerated. Beginning in the late 1980s, as the United States escalated its campaign against marijuana, American refugees from the drug war began moving to Amsterdam. Growers took with them their seeds and expertise, and this migration, matched with a Dutch genius for horticulture going back to the tulip craze, made Amsterdam, once again, the place to go if you cared deeply about one particular plant.

I went to Amsterdam to learn about the recent history of marijuana in America and to see—okay, and sample—what these gardeners had wrought in the years since my hasty retirement. I arrived in late November, at the time of the Cannabis Cup, an annual convention and harvest fair (sponsored by *High Times* magazine) that attracts many of the brighter lights in the field. American growers come to the Cup to do what gardeners always do when they gather in the off-season: swap seeds and stories and new techniques and show off their prize specimens. Some of the pioneers of modern marijuana growing were on hand, and I found that if I approached them as a fellow gardener, they were more than happy to share their experiences and knowledge.

Within a few days I had begun to piece together the story of how American gardeners, operating in the shadow of a ferocious drug war without benefit of professional training, had managed to transform "homegrown"—a derisive 1970s term for third-rate domestic marijuana—into what is today the most prized and expensive flower in the world.* But while the ingenuity and resourcefulness of growers had much to do with this success story, so did the ingenuity and resourcefulness of the plant itself. From the plant's perspective, the American drug war presented an opportunity to expand its range into North America, where it had never had much of a presence. (Except, that is, as hemp, a distinct, nonpsychoactive form of cannabis widely grown for its fibers before prohibition.) To succeed in North America, cannabis had to do two things: it had to prove it could gratify a human desire so brilliantly that people would take extraordinary risks to cultivate it, and it had to find the right combination of genes to adapt to a most peculiar and thoroughly artificial new environment. This is the story of how that happened.

:||:

Most of the marijuana smoked in America was grown in Mexico until the mid-1970s, when the Mexican government, at the behest of the United States, began spraying the crop with the herbicide paraquat. About the same time, the U.S. government began cracking down on pot smugglers. With foreign supplies contracting and

---

*Top-quality sinsemilla sells for upward of $500 an ounce, making cannabis America's leading cash crop.

the safety of Mexican marijuana in doubt, a large market for domestically grown marijuana suddenly opened up. In a sense, the rapid emergence of a domestic marijuana industry represents a triumph of protectionism.

In the beginning, domestic marijuana was grossly inferior to the imported product. Part of the problem was that most early growers did what I did: plant seeds picked out of pot that had been grown in tropical places. Invariably these were the seeds of *Cannabis sativa,* an equatorial species poorly adapted to life in the northern latitudes. *Sativa* can't withstand frost and, as I discovered, usually won't set flowers north of the thirtieth parallel. Working with such seeds, growers found it difficult to produce a high-quality domestic crop (and especially sinsemilla) outside places such as California and Hawaii.

The search was on for a type of marijuana that would flourish, and flower, farther north, and by the end of the decade, it had been found. American hippies traveling "the hashish trail" through Afghanistan returned with seeds of *Cannabis indica,* a stout, frost-tolerant species that had been grown for centuries by hashish producers in the mountains of central Asia. The species looks quite unlike the familiar marijuana plant (a distinct advantage to its early growers): it rarely grows taller than four or five feet (as compared to fifteen for the stateliest *sativas*), and its purplish green leaves are shorter and rounder than the long, slender fingers of *sativa. Indica* also proved to be exceptionally potent, although many people will tell you that its smoke is harsher and its high more physically debilitating than that of *sativa.* Even so, the introduction of *indica* to America proved a boon, since it allowed growers in all fifty states to cultivate sinsemilla for the first time. Some *indica*s will flower reliably as far north as Alaska.

Initially, *indica*s were grown by themselves. But enterprising growers soon discovered that by crossing the new species with *Cannabis sativa,* it was possible to produce vigorous hybrids that would combine the most desirable traits of each plant while downplaying its worst. The smoother taste and "clear, bell-like high" associated with the best equatorial *sativa*s, for example, could be combined with the superior potency and hardiness of an *indica.* The result was what Robert Connell Clarke, a marijuana botanist I met in Amsterdam, calls "the great revolution" in cannabis genetics.*

In a wave of innovative breeding performed around 1980, most of it by amateurs working in California and the Pacific Northwest, the modern American marijuana plant was born. Even today the *sativa* × *indica* hybrids developed during this period—including Northern Lights, Skunk #1, Big Bud, and California Orange—are regarded as the benchmarks of modern marijuana breeding; they remain the principal genetic lines with which most subsequent breeders have worked. Nowadays American cannabis genetics are widely regarded as the world's best; they are the basis of the thriving cannabis seed trade in Holland, as the American growers I met there were quick to point out. Yet without the Dutch to safeguard and disseminate these strains, the important genetic work done by

---

*Marijuana's genetic revolution recalls an earlier horticultural watershed: the introduction of the China rose (*R. chinensis*) to Europe in 1789, an event that made it possible for the first time to breed roses that would flower more than once a season. This ultimately led to the development of the ever-blooming hybrid tea rose. For both the rose and marijuana, human mobility coupled with human desire—for a rose that would rebloom in August; for sinsemilla that would grow in the north—led to the reunification of two distinct evolutionary lines of a plant that had diverged thousands of years before. In both cases, the introduction of a set of plant genes found halfway around the world created undreamed-of new possibilities.

American breeders would probably have been lost by now, scattered to the winds by the drug war.

‖:

Until the early 1980s, almost all the marijuana grown in America was grown outdoors: in the hills of California's Humboldt County, in the cornfields of the farm belt (cannabis and corn thrive under similar conditions), in backyards just about everywhere—and a lot more of it than anybody realized. In 1982 the Reagan administration was chagrined to discover that the amount of domestic marijuana being seized was actually a third higher than its official estimate of the *total* American crop. Shortly thereafter, the administration launched an ambitious nationwide program—enlisting local law enforcement agencies and, for the first time, the armed forces—to crush the domestic marijuana industry.

Though the government's campaign failed to eradicate marijuana farming, it did change the rules of the game, forcing both the plant and its growers to adapt: "The government pushed us all indoors," a grower from Indiana told me. And it was there, under the blazing metal halide lights, that *Cannabis sativa × indica* attained a kind of perfection.

The early indoor gardeners had basically sought to bring outdoor conditions and practices inside, growing full-size plants in soil under a regimen of light and nutrients designed more or less to mimic those found in nature. Very soon, however, growers discovered that nature was, if anything, holding back this particular plant, retarding its full potential. By judiciously manipulating the five main environmental factors under their control—water, nutrients, light, carbon dioxide levels, and heat—as well as the genetics of the plant, growers found that the marijuana plant, this remarkably obliging weed, could be made to perform wonders.

Most of the hybridizing needed to adapt cannabis to indoor conditions was done in the early 1980s by amateurs working in the Pacific Northwest. Cultivars with a high proportion of *indica* genes performed especially well indoors, it was found, and these were further bred and selected for small stature, high yield, early flowering, and increased potency. No one knew just what this plant was capable of, but by the end of the decade there were *sativa* × *indica* hybrids yielding flowers big as fists on dwarf plants no higher than your knee. During this period, cannabis genetics improved to the point where it was no longer unusual to find sinsemilla with concentrations of THC, marijuana's principal psychoactive compound, as high as 15 percent. (Before the crackdown on marijuana growers, THC levels in ordinary marijuana ranged from 2 to 3 percent, according to the DEA; for sinsemilla, 5 to 8 percent.) Nowadays THC levels upward of 20 percent are not unheard of.

The plant had adapted more brilliantly to its strange new environment than anyone could have expected. For cannabis, the drug war is what global warming will be for much of the rest of the plant world, a cataclysm that some species will turn into a great opportunity to expand their range. Cannabis has thrived on its taboo the way another plant might thrive in a particularly acid soil.

:|:

Along with the progress in genetics came rapid advances in technique. "Indoors," as one grower put it, "the gardener is Mother Nature, but even better." Growth rates and yields made large strides through the 1980s as growers discovered they could speed photosynthesis by supplying plants with all the nutrients, carbon dioxide, and light they could handle—vast amounts, as it turned out. (Cannabis is, after all, a weed.) Gardeners found that their plants could absorb hundreds of thousands of lumens—a blinding amount of light—twenty-four hours a day. Later on, by abruptly slashing their diet of light to twelve hours daily (and changing from metal halide to sodium lights, the frequency of which more closely mimics the autumn sun), growers could shock their plants into flowering before they were eight weeks old. With the right equipment, an indoor grower could create a utopia for his plants, an artificial habitat more perfect than any in nature, and his happy, happy weeds would respond.

These sedulous attentions would be wasted on male plants, which are worse than useless in sinsemilla production. As long as a female marijuana plant remains unpollinated, it will continue to produce new calyxes, steadily adding to the length of its flower. In this state of perpetual sexual frustration, the plant also continues to produce large quantities of THC-rich resins. But allow even a few grains of pollen to reach the plant's flowers, and the process abruptly stops: bud and resin production shuts down, the plant commences producing seeds—and the sinsemilla is ruined.

Growers who start their plants from seed rogue out the males as soon as they declare their gender, but since this doesn't happen until the plants mature, much time and space are wasted growing males. The solution was to plant clones instead of seeds—cuttings taken from established female "mother" plants. From the perspective of these fortunate females, the practice is an evolutionary boon: they get to multiply their genes without diluting them, as would be the case in sexual reproduction. (Whether cloning is such a boon for the species as a whole is, as the story of the apple suggests, much less certain.) Because these clones were genetically identical, the plants were guaranteed to be female. They also turned out to be biologically mature from the start, which meant that even a six- or eight-inch plant could now be forced to flower.

By 1987 all these various advances and techniques had coalesced into a state-of-the-art indoor growing regimen that came to be known as the Sea of Green: dozens of closely spaced and genetically identical plants grown from clones under high-intensity light. A Sea of Green garden consisting of a hundred clones, grown under a pair of thousand-watt lights in a space no bigger than a pool table, will yield three pounds of sinsemilla in two months' time.

:|:

Before I left Amsterdam I wanted to visit a modern marijuana garden, and on my last night an expatriate American grower I'd befriended agreed to show me his. For days I'd been fishing for an invitation, and I could see he was torn between the outlaw's professional discretion and the gardener's irrepressible desire to show off. In the end the gardener prevailed.

The garden was in a working-class suburb half an hour north of Amsterdam, and on the train the grower told me he'd chosen this particular town because it is home to a candy factory, a bakery, and a chemical plant. Marijuana plants, and *indica*s in particular, emit a strong, acrid odor; he was counting on the cacophony of smells produced by these three neighbors to cover the telltale stink of his plants.

When we came to the gardener's house, he showed me upstairs. At the far end of a dark, narrow, cluttered corridor, he flung open a tightly sealed door and I was hit squarely in the face first by a blast of white, white light, then by a stink so powerful it felt like a punch. Sweaty, vegetal, and sulfurous, the place might have been a locker room in the Amazon.

After my eyes adjusted to the light, I stepped into a windowless chamber not much bigger than a walk-in closet, crammed with electrical equipment, snaked with cables and plastic tubing, and completely sealed off from the world. More than half the room was taken up by the gardener's Sea of Green: a six-foot-square table invisible beneath a jungle of dark, serrated leaves oscillating gently in an artificial breeze. There were perhaps a hundred clones here, each barely a foot tall, yet already sending forth a thick finger of hairy calyxes, casting about vainly for a few grains of airborne pollen. A network of narrow plastic pipes supplied the plants with water, a tank of CO_2 sweetened their air, a ceramic heater warmed their roots at night, and four 600-watt sodium fixtures bathed them in a blaze of light for twelve hours of every day. The other twelve, they were sealed in perfect darkness. The briefest lapse of light, the gardener informed me gravely, would ruin the whole crop.

There was nothing of beauty in this garden. Should legalization ever come, no one is going to grow cannabis for the prettiness of its flowers, those hairy, sweaty-smelling, dandruffed clumps. There was also something bizarrely anomalous about this totalitarian hothouse, with its strict monoculture of genetically identical plants growing in lockstep—such ferocious Apollonian control in a garden ostensibly devoted to Dionysus.

Yet to a gardener there was much in this claustrophobic chamber to admire. I don't think I've ever seen plants that looked more enthusiastic, this despite the fact that they were being forced to grow in an utterly unnatural, even perverted manner: overbred, overfed, overstimulated, sped up, and pygmied all at once. "Happy to oblige!" the marijuana plants seemed to say, sucking up the CO_2, gorging themselves on the fertilizers, guzzling down the water, and throwing themselves at bulbs so hot and bright I finally had to look away. In exchange for a regimen of encouragement the likes of which few plants have ever known, these hundred eager demonic dwarves would oblige their gardener with three pounds of dried buds before the month was out—some $13,000 worth of flowers.

It was all more than a little mad, and very soon I was counting the minutes before I could politely make my exit and draw an ordinary breath. On the train back to Amsterdam, I tried to make some sense of this particular madness. It had a rather notorious local precedent, of course, an episode of equally intense involvement with a particular plant. During the tulipomania that briefly bewitched this city—the last time flowers traded for such fortunes—gardeners would exert themselves with a similar obsessiveness, rigging their precious plants with burglar alarms, deploying mirrors to multiply their blooms, and utterly failing to notice as their world shrank to the dimensions of a fevered dream.

One could argue that the fevered dream was the same then as now, and certainly visions of wealth have underwritten both the seventeenth-century tulip and twenty-first-century marijuana flower. Yet in the case of the tulip, by the end there was nothing *but* wealth to fuel the madness, and that surely is not the case with these other, uglier flowers. (The buds are homely, turdlike things, spangly with resin.) Tulipomania may have had as its spring the human desire for exotic visual pleasure, for beauty, but that didn't last. Beauty eventually gave way to status as the desire that drove otherwise rational people to navigate their lives by the polestar of this plant. And by the end pure financial speculation had hollowed out even that desire, so that no one noticed when the flowers were replaced by mere promises of themselves: the words on the paper of a futures contract.

The madness in the marijuana garden is of a different order. Though it too is abundantly watered by money, it remains deeply rooted in the human desire for pleasure—in whatever exactly it is that the chemicals produced in these flowers can do to a person's conscious experience. This desire must be an exceptionally powerful one—the passion and the price this flower commands have proven as much, as perhaps does the force of its taboo. Yet, for my part, I realized I didn't understand the first thing about that desire, not really. So what, exactly, *is* the knowledge held out by these plants, and why has it been so strenuously forbidden?

::

With the solitary exception of the Eskimos, there isn't a people on Earth who doesn't use psychoactive plants to effect a change in consciousness, and there probably never has been. As for the Eskimos, their exception only proves the rule: historically, Eskimos didn't use psychoactive plants because none of them will grow in the Arctic. (As soon as the white man introduced the Eskimo to fermented grain, he immediately joined the consciousness changers.) What this suggests is that the desire to alter one's experience of consciousness may be universal.

Nor is the desire limited to adults. Andrew Weil, who has written two valuable books treating consciousness changing "as a basic human activity," points out that even young children seek out altered states of awareness. They will spin until violently dizzy (thereby producing visual hallucinations), deliberately hyperventilate, throttle one another to the point of fainting, inhale any fumes they can find, and, on a daily basis, seek the rush of energy supplied by processed sugar (sugar being the child's plant drug of choice).

As the examples from childhood suggest, using drugs is not the only way to achieve altered states of consciousness. Activities as different as meditation, fasting, exercise, amusement park rides, horror movies, extreme sports, sensory or sleep deprivation, chanting, music, eating spicy foods, and taking extreme risks of all kinds have the power to change the texture of our mental experience to one degree or another. We may eventually discover that what psychoactive plants do to the brain closely resembles, at a biochemical level, the effects of these other activities.

Human cultures vary widely in the plants they use to gratify the desire for a change of mind, but all cultures (save the Eskimo) sanction at least one such plant and, just as invariably, strenuously forbid certain others. Along with the temptation seems to come the taboo. The reasons for drawing the bright line *here* and not *there* generally make more sense within the culture itself, rooted as they are in its values and traditions, than they do outside it. But the reasons cultures give for promoting one plant and forbidding another are remarkably fluid in both time and space; one culture's panacea is often another culture's panapathogen (root of all evil); think of the traditional role of alcohol in the Christian West as compared to the Islamic East. Indeed, one culture's panacea can, over time, transmogrify into that same culture's panapathogen, as happened to opiates in the West between the nineteenth and twentieth centuries.*

Historians can explain these shifts much better than scientists can, since they usually have less to do with the intrinsic nature of the various molecules involved than with the powers that cultures ascribe to them and the changing needs of those cultures. Cannabis in American culture has at various times held the power

---

*Tobacco smoking and coffee drinking were taboo in the West before the Industrial Revolution. The German historian Wolfgang Schivelbusch suggests that the two drugs became socially acceptable because they aided in industrialization's "reorientation of the human organism to the primacy of mental labor."

to foster violence (in the 1930s) and indolence (today): same molecule, opposite effect. Promoting certain plant drugs and forbidding others may just be something cultures do as a way of defining themselves or reinforcing their cohesion. It's hardly surprising that something as magical as a plant with the power to alter people's feelings and thoughts would inspire both fetishes and taboos.

:|:

What is harder to comprehend is why virtually all people, and more than a few animals, should have acquired such a desire in the first place. What good, from an evolutionary standpoint, could it do a creature to consume psychoactive plants? Possibly none at all: it's a fallacy to assume that whatever is is that way for a good Darwinian reason. Just because a desire or practice is widespread or universal doesn't necessarily mean it confers an evolutionary edge.

In fact, the human penchant for drugs may be the accidental by-product of two completely different adaptive behaviors. This at least is the theory Steven Pinker proposes in *How the Mind Works*. He points out that evolution has endowed the human brain with two (formerly) unrelated faculties: its superior problem-solving abilities and an internal system of chemical rewards, such that when a person does something especially useful or heroic the brain is washed in chemicals that make it feel good. Bring the first of these faculties to bear on the second, and you wind up with a creature who has figured out how to use plants to artificially trip the brain's reward system.

But doing so is not necessarily good for us. Ronald Siegel, the animal intoxication expert, has shown that animals who get high on plants tend to be more accident prone, more vulnerable to predators, and less likely to attend to their offspring. Intoxication is dangerous. But this only deepens the mystery: Why does the desire to alter consciousness remain powerful in the face of these perils? Or, put another way, why hasn't this desire simply died out, a casualty of Darwinian competition: the survival of the soberest?

The Greeks understood that the answer to most either/or questions about intoxicants (and a great many other of life's mysteries) is "Both." Dionysus's wine is both a scourge and a blessing. Used with care and in the proper context, many drug plants *do* confer advantages on the creatures that consume them—fiddling with one's brain chemistry can be very useful indeed. The relief of pain, a blessing of many psychoactive plants, is only the most obvious example. Plant stimulants, such as coffee, coca, and khat, help people to concentrate and work. Amazonian tribes take specific drugs to help them hunt, enhancing their endurance, eyesight, and strength. There are psychoactive plants that uncork inhibitions, quicken the sex drive, muffle or fire aggression, and smooth the waters of social life. Still others relieve stress, help people sleep or stay awake, and allow them to withstand misery or boredom. All these plants are, at least potentially, mental tools; people who know how to use them properly may be able to cope with everyday life better than those who don't.

⫶

These are the easy cases, though, the plants that merely inflect the prose of everyday life without rewriting it. "Transparent" is a term used to characterize drugs whose effects on consciousness are too subtle to interfere with one's ability to get through the day and fulfill one's obligations. Drugs such as coffee, tea, and tobacco in our culture, or coca and khat leaves in others, leave the user's space-time coordinates untouched. But what about the more powerful plants, the ones that *do* alter the experience of space and time in such a way as to take users out of everyday life—out of, even, themselves?

Cultures tend to be more wary of these plants, and for good reason: they pose a threat to the smooth workings of the social order. This may be why most complex, modern, secular societies have seen fit to forbid them. Even the cultures that endorse these plants cloak them in elaborate rules and rituals as a way of containing or disciplining their powers. So what are these powers, and what commends them—not only to adventurous individuals in all societies but, in some cases, to their societies as well? For many cultures have held these plants to be sacred.

:|:

No one has yet written the natural history of world religion, but we have some idea of the story such a book would tell. Among other things, it would force us to rethink the relation of matter and spirit—specifically, plant matter and human spirituality. For it would tell of how a select group of psychoactive plants and fungi (among them the peyote cactus, the *Amanita muscaria* and psilocybin mushrooms, the ergot fungus, the fermented grape, ayahuasca, and cannabis) were present at the creation of several of the world's religions. One of the world's earliest known religions was the cult of Soma, practiced by the ancient Indo-Europeans of central Asia; according to its sacred text, the Rig Veda, Soma was an intoxicant with the powers of a god. People worshiped the drug itself—which ethnobotanists now think was *Amanita muscaria,* the mushroom sometimes called fly agaric—as a path to divine knowledge.

Much the same process took place again and again all over the ancient world as people experimented, individually and in groups, with the power of plants to transcend the here and now and induce ecstasy—to take them elsewhere. What these peoples discovered was that certain plants or fungi (ethnobotanists call them "entheogens," meaning "the god within") opened a door onto another world. The images and words brought back from these journeys—visits with the souls of the dead and unborn, visions of the afterlife, answers to life's questions—were powerful enough to compel belief in a spirit world and, in some cases, to serve as the foundation of whole religions. Of course, plant drugs are not the only technologies of religious ecstasy; fasting, meditation, and hypnotic trances can achieve similar results. But often these techniques have been used to explore spiritual territory first blazed by the entheogens.

What a natural history of religion would show is that the human experience of the divine has deep roots in psychoactive plants and fungi. (Karl Marx may have gotten it backward when he called religion the opiate of the people.) This is not to diminish anyone's religious beliefs; to the contrary, that certain plants summon spiritual knowledge is precisely what many religious people have believed, and who's to say that belief is wrong? Psychoactive plants *are* bridges between the worlds of matter and spirit or, to update the vocabulary, chemistry and consciousness.

:|:

What a trick this is for a plant, to produce a chemical so mysterious in its effects on human consciousness that the plant itself becomes a sacrament, deserving of humankind's worshipful care and dissemination. Such was the fate of *Amanita muscaria* among the Indo-Europeans, peyote among the American Indians, cannabis among the Hindus, Scythians, and Thracians, wine among the Greeks* and early Christians.

---

*Judging from their descriptions of its effects, the Greeks probably fortified their wine with various psychoactive herbs; there's reason to think they also made religious use of ergot and *Amanita muscaria*.

In the same way the human desire for beauty and sweetness introduced into the world a new survival strategy for the plants that could gratify it, the human hunger for transcendence created new opportunities for another group of plants. No entheogenic plant or fungus ever set out to make molecules for the express purpose of inspiring visions in humans—combating pests is the far more likely motive. But the moment humans discovered what these molecules could do for them, this wholly inadvertent magic, the plants that made them suddenly had a brilliant new way to prosper. And from that moment on this is exactly what the plants with the strongest magic did.

⠿

Our desire for some form of transcendence of ordinary experience expresses itself not only in religion but in other endeavors as well, and these too have probably been more deeply influenced by psychoactive plants than we like to think. Who knows, we may need a natural history of literature and philosophy, or of discovery and invention, to go on the shelf with our natural history of religion. Or maybe what we need is just a single volume: a natural history of the imagination.

Somewhere in that volume we would surely find a chapter on the place of the opium poppy and cannabis in the romantic imagination. It's well known that many English romantic poets used opium, and several of the French romantics experimented with hashish soon after Napoleon's troops brought it back with them from Egypt. What's harder to know is precisely what role these psychoactive plants may have played in the revolution in human sensibility we call romanticism. The literary critic David Lenson, for one, believes it was crucial. He argues that Samuel Taylor Coleridge's notion of the imagination as a mental faculty that "dissolves, diffuses, dissipates, in order to re-create," an idea whose re-verberations in Western culture haven't yet been stilled, simply cannot be understood without reference to the change in consciousness wrought by opium.

"This notion of secondary or transforming imagination established a model of artistic creativity in the West that lasted from 1815 until the fall of Saigon," Lenson writes. "It is predicated on annihilating what Keats called 'weariness, fever and fret' (the world of fixed, dead objects) by just the sort of 'dissolution, diffusion and dissipation' that [moves the artist] toward the realms of accident, improvisation, and the unconscious." Not just romantic poetry, but modernism, surrealism, cubism, and jazz have all been nourished by Coleridge's idea of the transforming imagination—and that idea in turn was nourished by a psychoactive plant. "However criticism has tried to sanitize this process," Lenson writes, "we have to face the fact that some of our canonical poets and theorists, when apparently talking about imagination, are really talking about getting high."*

Curiously, the romantics at first believed it was their philosophical rather than poetical faculties that drugs would enhance. Thomas De Quincey felt that opium would give a philosopher "an inner eye and power of intuition for the vision and mysteries of our human nature." The nineteenth-century American writer Fitz Hugh Ludlow reported an important encounter with a philosopher of antiquity while under the spell of hashish. All of which makes me wonder: Is it possible that some of the philosophers of antiquity themselves had important encounters with magic plants?

---

*Sadie Plant, another literary critic, has argued that Coleridge's notion of the "suspension of disbelief" can also be traced to his use of opium.

This, at least, was my first thought upon learning that many of the important thinkers of classical Greece (including Plato, Aristotle, Socrates, Aeschylus, and Euripides) had participated in the "Mysteries of Eleusis." Nominally a harvest festival in honor of Demeter, the goddess of cultivated grains, the Mysteries were an ecstatic ritual during which participants consumed a powerful hallucinogenic potion. The precise recipe remains part of the mystery, but scholars speculate that the active ingredient was probably ergot, an alkaloid produced by a fungus (*Claviceps purpurea*) that infects cultivated grains and that closely resembles LSD in its chemical makeup and effects. Under the influence of this drug potion, the lights of classical civilization participated in a communal shamanic ritual of such mystery and transformative power that all who took part in it were sworn never to describe it. There is no way to know what, if anything, a philosopher or poet might have brought back from such a journey. But is it outlandish to ask whether such an experience might have helped inspire Plato's supernatural metaphysics—the belief that everything in our world has its true or ideal form in a second world beyond the reach of our senses?

One of the things certain drugs do to our perceptions is to distance or estrange the objects around us, aestheticizing the most commonplace things until they appear as ideal versions of themselves. Under the spell of cannabis "every object stands more clearly for all of its class," as David Lenson writes in *On Drugs*. "A cup 'looks like' the Platonic Idea of a cup, a landscape looks like a landscape painting, a hamburger stands for all the trillions of hamburgers ever served, and so forth." A psychoactive plant can open a door onto a world of archetypal forms, or so they can appear. Whether or not such a plant or fungus did this for Plato himself is of course impossible to ascertain, and somehow impious even to speculate on. But one could do worse, surely, searching for the spring of a metaphysics as visionary and strange as Plato's.

꞉ǀ꞉

The Platonic cup and the Coleridgean imagination are both "memes," to use a term coined by the British zoologist Richard Dawkins in his 1976 book, *The Selfish Gene*. A meme is simply a unit of memorable cultural information. It can be as small as a tune or a metaphor, as big as a philosophy or religious concept. Hell is a meme; so are the Pythagorean theorem, *A Hard Day's Night*, the wheel, *Hamlet*, pragmatism, harmony, "Where's the beef?," and of course the notion of the meme itself. Dawkins's theory is that memes are to cultural evolution what genes are to biological evolution. (Unlike genes, however, memes have no physical basis.) Memes are a culture's building blocks, passed down from brain to brain in a Darwinian process that leads, by trial and error, to cultural innovation and progress. The memes that prove themselves best adapted to their "environment"—that is, the ones that are most helpful for people to keep in their brains—are the ones most likely to survive and replicate and become widely regarded as good, true, or beautiful. Culture at any given moment is the "meme pool" in which we all swim—or rather, that swims through us.

Cultural change occurs whenever a new meme is introduced and catches on. It might be romanticism or double-entry bookkeeping, chaos theory or Pokémon. (Or the notion of memes itself, which seems to be catching on today.) So where in the world do new memes come from? Sometimes they spring full-blown from the brains of artists or scientists, advertising copywriters or teenagers. Often a process of mutation is involved in the creation of a new meme, in much the same way that mutations in the natural environment can lead to useful new genetic traits. Memes can mutate when they get combined in new ways, or when someone working with them makes a mistake—misreading or misinterpreting an old meme in such a way as to yield something new. For instance, besides being itself a new meme, Coleridge's transforming imagination has turned out to be an excellent technology for generating other new memes.

When I read Dawkins, it occurred to me that his theory suggested a useful way to think about the effects of psychoactive plants on culture—the critical role they've played at various junctures in the evolution of religion and music (think of jazz or rock improvisation), of poetry, philosophy, and the visual arts. What if these plant toxins function as a kind of cultural mutagen, not unlike the effect of radiation on the genome? They are, after all, chemicals with the power to alter mental constructs—to propose new metaphors, new ways of looking at things, and, occasionally, whole new mental constructs. Anyone who uses them knows they also generate plenty of mental errors; most such mistakes are useless or worse, but a few inevitably turn out to be the germs of new insights and metaphors. (And the better part of Western literature, if literary theorist Harold Bloom's idea of "creative misreading" is to be believed.) The molecules themselves don't add anything new to the stock of memes resident in a human brain, no more than radiation adds new genes. But surely the shifts in perception and breaks in mental habit they provoke are among the methods, and models, we have of imaginatively transforming mental and cultural givens—for mutating our inherited memes.

∷

At the risk of discrediting my own idea, I want to acknowledge that it owes a debt—how large I can't say—to a psychoactive plant. The notion that drugs might function as cultural mutagens occurred to me while reading *The Selfish Gene* while high on marijuana, which may or may not be an advisable thing to do. But whatever its value, it's at least a fresh idea (itself a kind of mutation of Dawkins's meme idea), and I seriously doubt it would have occurred to me had I not smoked a little pot the evening I was reading Dawkins. (I wish I could say the same about the earlier speculation on Plato, but I'm afraid I was straight as a post for that one.)

I know, I said that I didn't much like smoking pot. But research is research, and besides, my personal relationship to cannabis underwent a sea change while I was in Amsterdam. I'd heard so much about the improvements made to marijuana that I felt I had to give it another try, and I promptly discovered that this pot, at least, left me feeling neither stupid nor paranoid.

The nonstupid part can, I think, be accounted for by advances in cannabis breeding that make it possible to develop strains eliciting distinctly different mental effects. At the top end of the market this has led to a connoisseurship of cannabis—not just of its taste or aroma, but of the specific psychological texture of its high. Some strains (typically those with a higher proportion of *indica* genes) are narcotic in their effects, tending to stupefy. Others (often the ones with more *sativa* genes) leave the mind clear and fluent and the body unimpaired. Some of the growers I met spoke in terms of "white-collar" and "blue-collar" pot. The strains I found personally sympathetic were stimulating and, evidently, conducive to mental speculation.

As for the nonparanoid part, remember that I was in a country where one can smoke marijuana openly and without fear. The effect of the American drug war on the experience of smoking marijuana—a drug notoriously susceptible to the power of suggestion—cannot be overestimated. Writing in *The Atlantic Monthly* in 1966 about the intellectual "uses" of marijuana (now, *there's* a topic that's moved beyond the pale; these days one may speak of marijuana's medicinal uses, perhaps, but *intellectual?*), Allen Ginsberg suggested that the negative feelings marijuana sometimes provokes, such as anxiety, fear, and paranoia, are "traceable to the effects on consciousness not of the narcotic but of the law." Researchers speak of "set and setting" as crucial factors shaping one's experience of any drug, and marijuana in particular almost unfailingly fulfills one's expectation of it, for better and worse. Lenson calls it "the great yea-sayer, supporting whatever is going on anyway, and introducing little or nothing of its own." In my experience, cannabis can't reliably be used to change one's mood, only to intensify it. Smoking in a comfortable coffee shop with a dozen other people doing the same thing, I had no reason to feel paranoid, which is probably why I didn't.

Taking account of this phenomenon, Andrew Weil describes marijuana as an "active placebo." He contends that cannabis does not itself create but merely triggers the mental state we identify as "being high." The very same mental state, minus the "physiological noise" of the drug itself, can be triggered in other ways, such as meditation or breathing exercises. Weil believes it is an error of modern materialistic thinking to believe (as both drug users and drug researchers invariably do) that the "high" smokers experience is somehow a product of the plant itself (or THC), rather than a creation of the mind—prompted, perhaps, but *sui generis*.

The truth of the matter is probably where it usually is, somewhere in the middle. Certainly the psychological experience of marijuana is far too varied, not only from person to person but from time to time, to be explained purely in terms of a chemical. At the same time, the chemistry of this *particular* plant surely has something specific to do with, say, the novel perceptions of Cézanne's pictorial space that Ginsberg describes in his *Atlantic* essay, the religious insights brought back by shamans, or even my own vagrant speculations on mutating memes. Opium would probably induce different kinds of thoughts in the same brains. We assume that there is some sort of cause-and-effect relationship between molecule and mind, but what it is no one really knows.

As the sorcerers, shamans, and alchemists who used them understood, psychoactive plants stand on the threshold of matter and spirit, at the point where simple distinctions between the two no longer hold. Consciousness is what we're talking about here, of course, and consciousness is precisely the frontier where our materialistic understanding of the brain stops—at least for the time being, but possibly forever. What's interesting about a plant like marijuana is that it takes us right up to that frontier and may have something to teach us about what lies on the other side. We tend to smile indulgently at poets like Allen Ginsberg for believing that cannabis is a useful tool for exploring consciousness. But it turns out they may be right.

⫶⫶

In the mid-1960s, an Israeli neuroscientist named Raphael Mechoulam identified the chemical compound responsible for the psychoactive effects of marijuana: delta-9-tetrahydrocannabinol, or THC, a molecule with a structure unlike any found in nature before or since. For years Mechoulam had been intrigued by the ancient history of cannabis as a medicine (a panacea in many cultures until its prohibition in the 1930s, it has been used to treat pain, convulsions, nausea, glaucoma, neuralgia, asthma, cramps, migraine, insomnia, and depression) and decided it might be worthwhile to isolate the plant's active ingredient. But it was the popularity of marijuana as a recreational drug in the sixties, and the attendant official worries, that freed up the resources to underwrite this kind of work—and a great deal of other cannabinoid research that, taken together, has yielded more knowledge about the workings of the human brain than anyone could have guessed.

In 1988 Allyn Howlett, a researcher at the St. Louis University Medical School, discovered a specific receptor for THC in the brain—a type of nerve cell that THC binds to like a molecular key in a lock, causing it to activate. Receptor cells form part of a neuronal network; the brain systems involving dopamine, serotonin, and the endorphins are three such networks. When a cell in a network is activated by its chemical key, it responds by doing a variety of things: sending a chemical signal to other cells, switching a gene on or off, or becoming more or less active. Depending on the network involved, this process can trigger cognitive, behavioral, or physiological changes. Howlett's discovery pointed to the existence of a new network in the brain.

The cannabinoid receptors Howlett found showed up in vast numbers all over the brain (as well as in the immune and reproductive systems), though they were clustered in regions responsible for the mental processes that marijuana is known to alter: the cerebral cortex (the locus of higher-order thought), the hippocampus (memory), the basal ganglia (movement), and the amygdala (emotions). Curiously, the one neurological address where cannabinoid receptors *didn't* show up was in the brain stem, which regulates involuntary functions such as circulation and respiration. This might explain the remarkably low toxicity of cannabis and the fact that no one is known to have ever died from an overdose.

On the assumption that the human brain would not have

evolved a special structure for the express purpose of getting it-self high on marijuana, researchers hypothesized that the brain must manufacture its own THC-like chemical for some as-yet-unknown purpose. (The scientific paradigm at work here was the endorphin system, which is tripped by opiates from plants as well as endorphins produced in the brain.) In 1992, some thirty years after his discovery of THC, Raphael Mechoulam (working with a collaborator, William Devane) found it: the brain's own endoge-nous cannabinoid. He named it "anandamide," from the Sanskrit word for "inner bliss."

Someday soon Mechoulam and Howlett will almost surely re-ceive the Nobel Prize, for their discoveries opened a new branch of neuroscience that promises to revolutionize our understanding of the brain and lead to a whole new class of drugs. Following on their work, neuroscientists are now busy trying to figure out ex-actly how the cannabinoid network works—and why we should have one in the first place.

I put that question to Mechoulam and Howlett and several of their colleagues in cannabinoid research, and their answers, while speculative, are richly suggestive. The cannabinoid network is un-usually complex and varied in its functions, I learned, in part be-cause it seems to modulate the action of other neurotransmitters, such as serotonin, dopamine, and the endorphins. When I asked Howlett what the purpose of such a network might be, she began her answer by listing some of the various direct and indirect ef-fects of cannabinoids: pain relief, loss of short-term memory, se-dation, and mild cognitive impairment.

"All of which is *exactly* what Adam and Eve would want after being thrown out of Eden. You couldn't design a more perfect drug for getting Eve through the pain of childbirth or helping Adam endure a life of physical toil." She noted that cannabinoid receptors had been found in the uterus, of all places, and speculated that anandamide may not only dull the pain of childbirth but help women forget it later. (The sensation of pain is, curiously, one of the hardest to summon from memory.) Howlett speculated that the human cannabinoid system evolved to help us endure (and selectively forget) the routine slings and arrows of life "so that we can get up in the morning and do it all over again." It is the brain's own drug for coping with the human condition.

For his part, Raphael Mechoulam believes that the cannabinoid network is involved in regulating several different biological processes, including pain management, memory formation, appetite, the coordination of movement, and, perhaps most intriguingly, emotion. "We know next to nothing about the biochemistry of emotion," Mechoulam points out, but he thinks we'll eventually discover that cannabinoids are involved in the process by which the brain "translates objective reality into subjective emotions."

"If I see my grandson rushing to meet me, I feel happy. How do I translate biochemically the objective reality of a grandson rushing toward me into the subjective change in my emotions?" The brain's cannabinoids could be the missing link.

∷

So what are the odds that a molecule produced by a flower out in the world—by a weedy plant native to central Asia—would turn out to hold the precise key required to unlock the neurological mechanism governing these aspects of human consciousness? There is something miraculous about such a correspondence between nature and mind, yet it must have a logical explanation. A plant does not go to the expense of making (and continuing to make) such a unique and complex molecule if it doesn't do the plant some evolutionary good. So why does cannabis produce

THC? No one knows for sure, but botanists offer several competing theories, and most of them have nothing to do with getting people high—at least not at the plant's beginnings.

The purpose of THC could be to protect cannabis plants from ultraviolet radiation; it seems that the higher the altitude at which cannabis grows, the more THC it produces. THC also exhibits antibiotic properties, suggesting a role in protecting cannabis from disease. Last, it's possible that THC gives the cannabis plant a sophisticated defense against pests. Cannabinoid receptors have been found in animals as primitive as the hydra, and researchers expect to find them in insects. Conceivably, cannabis produces THC to discombobulate the insects (and higher herbivores) that prey on the plant; it might make a bug (or a buck or a rabbit) forget what it's doing or where in the world it last saw that tasty plant. But whatever THC's purpose, it's unlikely that, as Raphael Mechoulam put it, "a plant would produce a compound so that a kid in San Francisco can get high."

Or is it? Robert Connell Clarke, the marijuana botanist I met in Amsterdam, doesn't think that notion is quite as far-fetched as Mechoulam makes it sound. He finds most of the defense theories inadequate and concludes that "the most obvious evolutionary advantage THC conferred on *Cannabis* was the psychoactive properties, which attracted human attention and caused the plant to be spread around the world."

Of course, Mechoulam and Clarke could both be right. Whatever THC's original purpose may have been, as soon as a certain primate with a gift for experiment and horticulture stumbled on its psychoactive properties, the plant's evolution embarked on a new trajectory, guided from then on by that primate and his desires. The cannabis flowers that gave humans the most pleasure, or strongest medicine, were now the ones that produced the most offspring. What may have started out as a biochemical accident became the plant's coevolutionary destiny, or at least *one* of its destinies, under domestication.

Ma, the ancient Chinese character for "hemp," depicts a male and a female plant under a roof—cannabis inside the house of human culture. Cannabis was one of the earliest plants to be domesticated (probably for fiber first, then later as a drug); it has been coevolving with humankind for more than ten thousand years, to the point where the aboriginal form of the plant may no longer exist. By now cannabis is as much the product of human desire as a Bourbon rose, and we have scant idea what the plant might have been like before it linked its destiny to our own.

But what is so unusual about cannabis's coevolution (compared to that of the rose, say, or the apple) is that it followed two such divergent paths down to our time, each reflecting the influence of a completely different human desire. Along the first path (which appears to have begun in ancient China and moved west toward northern Europe, then on to the Americas), the plant was selected by people for the strength and length of its fibers. (Up until the last century, hemp was one of humankind's main sources of paper and cloth.) Along the other path (which began somewhere in central Asia and moved down through India, then into Africa, and from there across to the Americas with the slaves and up to Europe with Napoleon's army), cannabis was selected for its psychoactive and medicinal powers. Ten thousand years later, hemp and cannabis are as different as night and day: hemp produces negligible amounts of THC and cannabis a worthless fiber. (In the eyes of the U.S. government, however, there is still only one plant, so that the taboo on the drug plant has, pointlessly, doomed the fiber.) It is hard to conceive of a domesticated plant more plastic than cannabis, a single species answering to two such different desires, the first more or less spiritual in nature and the other, quite literally, material.

∷

The scientists I talked to had a lot to say about the descent and biochemistry of cannabis, but about the plant's effects on our experience of consciousness they were all but silent. What I wanted to know is, What exactly does it mean, biologically, to say a person is "high"? When I put this question to Allyn Howlett, her answer consisted of two rather parched words: "cognitive dysfunction." *Cognitive dysfunction?* Okay, but isn't that a little like saying that having sex elevates one's pulse? It's perfectly true as far as it goes, but it doesn't get you any closer to the heart of the matter—or to the desire. John Morgan, a pharmacologist who has written widely about marijuana, points out that "we don't yet understand consciousness scientifically, so how can we hope to explain changes in consciousness scientifically?" Mechoulam replied to my questions about what it means biochemically to be high simply by saying, "I am afraid we have to leave these questions still to the poets."

So there it seemed the neuroscientists had stranded me, all on my unscientific own with a dime bag and the dubious company of poets such as Allen Ginsberg and Charles Baudelaire, Fitz Hugh Ludlow and (yikes!) Carl Sagan—but Carl Sagan wearing his goofiest nonscientific hat. You see, I'd discovered that in 1971 Sagan had anonymously published an earnest, marvelous account of his experiences with pot, which he credited with "devastating insights" about the nature of life.*

---

*"There is a myth about such highs," Sagan wrote; "the user has an illusion of great insight, but it does not survive scrutiny in the morning. I am convinced that this is an error, and that the devastating insights achieved while high are real in-

Yet as I proceeded with my literary and phenomenological investigations of the pot experience, I soon realized I had gotten something valuable from the scientists after all. They had inadvertently pointed me in the direction of a deeper understanding of what it is that cannabis does to human consciousness and what, possibly, it has to teach us about it. In fact, Howlett was probably right, if inelegant, in her simple formulation, because I've come to think that a "cognitive dysfunction" of a very special kind does in fact lie at the heart of it. Let me try to explain.

The scientists I spoke to were unanimous in citing short-term memory loss as one of the key neurological effects of the cannabinoids. In their own way, so were the "poets" who tried to describe the experience of cannabis intoxication. All talk about the difficulty of reconstructing what happened mere seconds ago and what a Herculean challenge it becomes to follow the thread of a conversation (or a passage of prose) when one's short-term memory isn't operating normally.

Yet the scientists said that the THC in cannabis is only mimicking the actions of the brain's own cannabinoids. What a curious thing this is for a brain to do, to manufacture a chemical that interferes with its own ability to make memories—and not just

---

sights; the main problem is putting these insights in a form acceptable to the quite different self that we are when we're down the next day. . . . If I find in the morning a message from myself the night before informing me that there is a world around us which we barely sense, or that we can become one with the universe, or even that certain politicians are desperately frightened men, I may tend to disbelieve; but when I'm high I know about this disbelief. And so I have a tape in which I exhort myself to take such remarks seriously. I say, 'Listen closely, you son-ofabitch of the morning! This stuff is real!' " Sagan's essay, attributed to "Mr. X," appears in *Marihuana Reconsidered*, by Lester Grinspoon. After Sagan's death in 1996, Grinspoon revealed Mr. X's identity.

memories of pain, either. So I e-mailed Raphael Mechoulam to ask him why he thought the brain might secrete a chemical that has such an undesirable effect.

Don't be so sure that forgetting is undesirable, he suggested. "Do you really want to remember all the faces you saw on the New York City subway this morning?"

Mechoulam's somewhat oblique comment helped me begin to appreciate that forgetting is vastly underrated as a mental operation—indeed, that it *is* a mental operation, rather than, as I'd always assumed, strictly the breakdown of one. Yes, forgetting can be a curse, especially as we age. But forgetting is also one of the more important things healthy brains do, almost as important as remembering. Think how quickly the sheer volume and multiplicity of sensory information we receive every waking minute would overwhelm our consciousness if we couldn't quickly forget a great deal more of it than we remember.

At any given moment, my senses present to my consciousness—this perceiving "I"—a blizzard of data no human mind can completely absorb. To illustrate the point, let me try to capture here a few drops of this perceptual cataract, preserve one cross section of the routinely forgotten. Right now my eyes, even without moving, offer the following: directly in front of me, the words I'm typing on a computer screen along with its blue background and tumble of icons. Peripherally, there's the blond wood grain of my desk, a mouse pad (printed with words and images), a CD spinning red in its little window, two bookshelves crammed with a couple of dozen spines I could easily read but don't, a gray plastic heater grate, a blue folder (entitled "Pot clips") stuck into a standing file at an annoying angle, two hands with an unspecified number of flying fingers (Band-Aid on one hand, glint of gold on the other), one jeans-clad lap, two green-sweatered wrists, a window (its green muntins framing a boulder with lichens, dozens of trees, hundreds of branches, millions of leaves), and, drawing a soft border around 90 percent of this visual field, the metal frames of my eyeglasses.

And that's just my eyes. My sense of touch meanwhile presents to my attention a low background drone of shoulder ache, a slight burning sensation in the tip of my right middle finger (where it was cut the other day), and the cool rush of air through my nostrils. Taste? Black tea and bergamot (Earl Grey), slightly briny breakfast residue on tongue (smoked salmon). Soundtrack: Red Hot Chili Peppers in the foreground, backed by heater whoosh on the right, computer cooling fan whoosh on the lower left, mouse clicks, keyboard clatter, creak-crack of those knuckle-like things deep in the neck when I cant my head to one side; and then, outside, a scatter of birdsong, methodical drips on the roof, and the slow sky tear of a propeller plane. Smell: Lemon Pledge, mixed with woodsy damp. I won't even try to catalog the numberless errant thoughts presently nipping around the writing of this paragraph like a flittering school of fish. (Or maybe I will: second thoughts and misgivings arriving in waves, shoving crowds of alternative words and grammatical constructions, shimmering lunch options, small black holes of consciousness from which I try to fish out metaphors, a clamoring handful of to-dos, a spongy awareness of the time till lunch, and so on, and so on.)

"If we could hear the squirrel's heartbeat, the sound of the grass growing, we should die of that roar," George Eliot once wrote. Our mental health depends on a mechanism for editing the moment-by-moment ocean of sensory data flowing into our consciousness down to a manageable trickle of the noticed and remembered. The cannabinoid network appears to be part of that mechanism, vigilantly sifting the vast chaff of sense impression from the kernels of perception we need to remember if we're to get through the day and get done what needs to be done.* Much depends on forgetting.

The THC in marijuana and the brain's endogenous cannabinoids work in much the same way, but THC is far stronger and more persistent than anandamide, which, like most neurotransmitters, is designed to break down very soon after its release. (Chocolate, of all things, seems to slow this process, which might account for its own subtle mood-altering properties.) What this suggests is that smoking marijuana may overstimulate the brain's built-in forgetting faculty, exaggerating its normal operations.

This is no small thing. Indeed, I would venture that, more than any other single quality, it is the relentless moment-by-moment forgetting, this draining of the pool of sense impression almost as quickly as it fills, that gives the experience of consciousness under marijuana its peculiar texture. It helps account for the sharpening of sensory perceptions, for the aura of profundity in which cannabis bathes the most ordinary insights, and, perhaps most important of all, for the sense that time has slowed or even stopped. For it is only by forgetting that we ever really drop the thread of time and approach the experience of living in the present moment, so elusive in ordinary hours. And the wonder of *that* experience, perhaps more than any other, seems to be at the very heart of the human desire to change consciousness, whether by means of drugs or any other technique.

---

*Mechoulam thinks we'll eventually find a neurotransmitter that does for remembering what the cannabinoids do for forgetting, and that the push-and-pull interaction of these two chemicals together determines what is filed in memory and what is thrown out.

:|:

"Consider the cattle, grazing as they pass you by," Friedrich Nietzsche begins a brilliant, somewhat eccentric 1876 essay he called "The Uses and Disadvantages of History for Life." "They do not know what is meant by yesterday or today, they leap about, eat, rest, digest, leap about again, and so from morn till night and from day to day, fettered to the moment and its pleasure or displeasure, and thus neither melancholy nor bored. . . .

"A human being may well ask an animal: 'Why do you not speak to me of your happiness but only stand and gaze at me?' The animal would like to answer, and say, 'The reason is I always forget what I was going to say'—but then he forgot this answer too, and stayed silent."

The first part of Nietzsche's essay is a moving and occasionally hilarious paean to the virtues of forgetting, which he maintains is a prerequisite to human happiness, mental health, and action. Without dismissing the value of memory or history, he argues (much like Emerson and Thoreau) that we spend altogether too much of our energy laboring in the shadows of the past—under the stultifying weight of convention, precedent, received wisdom, and neurosis. Like the American transcendentalists, Nietzsche believes that our personal and collective inheritance stands in the way of our enjoyment of life and accomplishment of anything original.

"Cheerfulness, the good conscience, the joyful deed, confidence in the future—all of them depend . . . on one's being just as able to forget at the right time as to remember." He admonishes us to cast off "the great and ever-greater pressure of what is past" and live instead rather more like the child (or the cow) that "plays in blissful blindness between the hedges of past and future." Nietzsche acknowledges that there are perils to inhabiting the present (one is liable to "falsely suppose all his experiences are original to him"), but any loss in knowingness or sophistication is more than made up for by the gain in vigor.

For Nietzsche the "art and power of forgetting" consist in a kind of radical editing or blocking out of consciousness everything that doesn't serve the present purpose. A man seized by a "vehement passion" or great idea will be blind and deaf to all except that passion or idea. Everything he does perceive, however, he will perceive as he has never perceived anything before: "All is so palpable, close, highly colored, resounding, as though he apprehended it with all his senses at once."

What Nietzsche is describing is a kind of transcendence—a mental state of complete and utter absorption well known to artists, athletes, gamblers, musicians, dancers, soldiers in battle, mystics, meditators, and the devout during prayer. Something very like it can occur during sex, too, or while under the influence of certain drugs. It is a state that depends for its effect on losing oneself in the moment, usually by training a powerful, depthless concentration on One Big Thing. (Or, in the Eastern tradition, One Big Nothing.) If you imagine consciousness as a kind of lens through which we perceive the world, the drastic constricting of its field of vision seems to heighten the vividness of whatever remains in the circle of perception, while everything else (including our awareness of the lens itself) simply falls away.

Some of our greatest happinesses arrive in such moments, during which we feel as though we've sprung free from the tyranny of time—clock time, of course, but also historical and psychological time, and sometimes even mortality. Not that this state of mind doesn't have its drawbacks; to name one, other people cease to matter. Yet this thoroughgoing absorption in the present is (as both Eastern and Western religious traditions tell us) as close as we mortals ever get to an experience of eternity. Boethius, the sixth-century Neoplatonist, said the goal of our spiritual striving was "to hold and possess the whole fullness of life in one moment, here and now, past and present and to come." Likewise in the Eastern tradition: "Awakening to this present instant," a Zen master has written, "we realize the infinite is in the finite of each instant." Yet we can't get there from here without first forgetting.

::

I am not by nature one of the world's great noticers. Unless I make a conscious effort, I won't notice what color your shirt is, the song playing on the radio, or whether you put one sugar in your coffee or two. When I'm working as a reporter I have to hector myself continually to mark the details: checked shirt, two sugars, Van Morrison. Why this should be so, I have no idea, except that I am literally absentminded, prone to be thinking about something else, something past, when I am ostensibly having a fresh experience. Almost always, my attention can't wait to beat a retreat from the here and now to the abstract, frog-jumping from the data of the senses to conclusions.

Actually, it's worse than that. Very often the conclusions or concepts come first, allowing me to dispense with the sensory data altogether or to notice in it only what fits. It's a form of impatience with lived life, and though it might appear to be a symptom of an active mind, I suspect it's really a form of laziness. My lawyer father, once complimented on his ability to see ahead three or four moves in a negotiation, explained that the reason he liked to jump to conclusions was so he could get there early and rest. I'm the same way in my negotiations with reality.

Though I suspect that what I have is only an acute case of an attention disorder that is more or less universal. Seeing, hearing, smelling, feeling, or tasting things as they "really are" is always difficult if not impossible (in part because doing so would overwhelm us, as George Eliot understood), so we perceive each multisensory moment through a protective screen of ideas, past experiences, or expectations. "Nature always wears the colors of the spirit," Emerson wrote, by which he meant we never see the world plainly, only through the filter of prior concepts or metaphors. ("Colors," in classical rhetoric, are tropes.) In my case this filter is so fine (or is it thick?) that a lot of the details and textures of reality simply never get through. It's a habit of mind I sorely wish I could break, since it keeps me from enjoying the pleasures of the senses and the moment, pleasures that, at least in the abstract, I prize above all others. But right there you see the problem: *in the abstract.*

All those who write about cannabis's effect on consciousness speak of the changes in perception they experience, and specifically of an intensification of all the senses. Common foods taste better, familiar music is suddenly sublime, sexual touch revelatory. Scientists who've studied the phenomenon can find no quantifiable change in the visual, auditory, or tactile acuity of subjects high on marijuana, yet these people invariably report seeing, and hearing, and tasting things with a new keenness, as if with fresh eyes and ears and taste buds.

You know how it goes, this italicization of experience, this seemingly virginal *noticing* of the sensate world. You've heard that song a thousand times before, but now you suddenly *hear* it in all its soul-piercing beauty, the sweet bottomless poignancy of the guitar line like a revelation, and for the first time you can understand, *really* understand, just what Jerry Garcia meant by every note, his unhurried cheerful-baleful improvisation piping something very near the meaning of life directly into *your* mind.

Or that exceptionally delicious spoonful of vanilla ice cream—*ice cream!*—parting the drab curtains of the quotidian to reveal, what?—the heartrendingly sweet significance of *cream,* yes, bearing us all the way back to the breast. Not to mention the never-before-adequately-appreciated wonder of: *vanilla.* How astonishing is it that we happen to inhabit a universe in which this quality of vanilla-ness—this *bean!*—happens also to reside? How easily it could have been otherwise, and just where would we be (where would *chocolate* be?) without that singular irreplaceable note, that middle C on the Scale of Archetypal Flavors? (*Paging Dr. Plato!*) For the first time in your journey on this planet you are fully appreciating *Vanilla* in all its italicized and capitalized significance. Until, that is, the next epiphany comes along (*Chairs! People thinking in other languages! Carbonated water!*) and the one about ice cream is blown away like a leaf on the breeze of free association.

Nothing is easier to make fun of than these pot-sponsored perceptions, long the broad butt of jokes about marijuana. But I'm not prepared to concede that these epiphanies are as empty or false as they usually appear in the cold light of the next day. In fact, I'm tempted to agree with Carl Sagan, who was convinced that marijuana's morning-after problem is not a question of self-deception so much as a failure to communicate—to put "these insights in a form acceptable to the quite different self that we are when we're down the next day." We simply don't have the words to convey the force of these perceptions to our straight selves, perhaps because they are the kinds of perceptions that precede words. They may well be banal, but that doesn't mean they aren't also at the same time profound.

Marijuana dissolves this apparent contradiction, and it does so by making us temporarily forget most of the baggage we usually bring to our perception of something like ice cream, our acquired sense of its familiarity and banality. For what is a sense of the banality of something if not a defense against the overwhelming (or at least whelming) power of that thing experienced freshly? Banality depends on memory, as do irony and abstraction and boredom, three other defenses the educated mind deploys against experience so that it can get through the day without being continually, exhaustingly astonished.

It is by temporarily mislaying much of what we already know (or think we know) that cannabis restores a kind of innocence to our perceptions of the world, and innocence in adults will always flirt with embarrassment. The cannabinoids are molecules with the power to make romantics and transcendentalists of us all. By disabling our moment-by-moment memory, which is ever pulling us off the astounding frontier of the present and throwing us back onto the mapped byways of the past, the cannabinoids open a space for something nearer to direct experience. By the grace of this forgetting, we temporarily shelve our inherited ways of looking and see things as if for the first time, so that even something as ordinary as ice cream becomes *Ice cream!*

There is another word for this extremist noticing—this sense of first sight unencumbered by knowingness, by the already-been-theres and seen-thats of the adult mind—and that word, of course, is *wonder.*

∷

Memory is the enemy of wonder, which abides nowhere else but in the present. This is why, unless you are a child, wonder depends on forgetting—on a process, that is, of subtraction. Ordinarily we think of drug experiences as additive—it's often said that drugs "distort" normal perceptions and augment the data of the senses (adding hallucinations, say), but it may be that the very opposite is true—that they work by subtracting some of the filters that consciousness normally interposes between us and the world.

This, at least, was Aldous Huxley's conclusion in *The Doors of Perception*, his 1954 account of his experiments with mescaline. In Huxley's view, the drug—which is derived from peyote, the flower of a desert cactus—disables what he called "the reducing valve" of consciousness, his name for the conscious mind's everyday editing faculty. The reducing valve keeps us from being crushed under the "pressure of reality," but it accomplishes this at a price, for the mechanism prevents us from ever seeing reality as it really is. The insight of mystics and artists flows from their special ability to switch off the mind's reducing valve. I'm not sure any of us ever perceives reality "as it really is" (how would one know?), but Huxley is persuasive in depicting wonder as what happens when we succeed in suspending our customary verbal and conceptual ways of seeing. (He writes with a wacky earnestness about the beauty of fabric folds, a garden chair, and a vase of flowers: "I was seeing what Adam had seen on the morning of his creation—the miracle, moment by moment, of naked existence.")

I think I understand Huxley's reducing valve of consciousness, though in my own experience the mechanism looks a little different. I picture ordinary consciousness more as a funnel or, even better, as the cinched waist of an hourglass. In this metaphor the mind's eye stands poised between time past and time to come, determining which of the innumerable grains of sensory experience will pass through the narrow aperture of the present and enter into memory. I know, there are some problems with this metaphor, the main one being that all the sand eventually gets to the bottom of an hourglass, whereas most of the grains of experience never make it past our regard. But the metaphor at least gets at the notion that the principal work of consciousness is eliminative and defensive, maintaining perceptual order to keep us from being overwhelmed.

So what happens under the influence of drugs or, for that matter, inspiration? In Huxley's metaphor, the reducing valve is opened wide to admit more of experience. This seems about right, though I'd qualify it by suggesting (as Huxley's own examples do) that the effect of altered consciousness is to admit a whole lot more information about a much smaller increment of experience. "The folds of my gray flannel trousers were charged with 'is-ness,' " Huxley tells us, before dilating on Botticelli draperies and the "Allness and Infinity of folded cloth." The usual process by which the grains of perception pass us by slows way down, to the point where the conscious I can behold each grain in its turn, scrupulously examining it from every conceivable angle (sometimes from more angles than it even has), until all there is is the still point at the hourglass's waist, where time itself appears to pause.

::|::

But is this wonder the real thing? At first glance, it wouldn't seem to be: a transcendence that's chemically induced must surely be fake. *Artificial Paradises* was what Charles Baudelaire called his 1860 book about his experiences with hashish, and that sounds about right. Yet what if it turns out that the neurochemistry of transcendence is no different whether you smoke marijuana, meditate, or enter a hypnotic trance by way of chanting, fasting, or prayer? What if in every one of these endeavors, the brain is simply prompted to produce large quantities of cannabinoids, thereby suspending short-term memory and allowing us to experience the present deeply? There are many technologies for changing the brain's chemistry; drugs may simply be the most direct. (This doesn't necessarily make drugs a *better* technology for changing consciousness—indeed, the toxic side effects of so many of them suggest that the opposite is true.) From a brain's point of view, the distinction between a natural and an artificial high may be meaningless.

Aldous Huxley did his best to argue us out of the view that a chemically conditioned spiritual experience is false—and he did so long before we knew anything about cannabinoid or opioid receptor networks. "In one way or another, *all* our experiences are chemically conditioned, and if we imagine that some of them are purely 'spiritual,' purely 'intellectual,' purely 'aesthetic,' it is merely because we have never troubled to investigate the internal chemical environment at the moment of their occurrence." He points out that mystics have always worked systematically to modify their brain chemistry, whether through fasting, self-flagellation, sleeplessness, hypnotic movement, or chanting.* The brain can be made to drug itself, as seems to happen with certain placebos. We don't merely imagine that the placebo antidepressant is working to lift our sadness or worry— the brain is actually producing extra serotonin in response to the mental prompt of swallowing a pill containing nothing but sugar and belief. What all this suggests is that the workings of consciousness are both more and less materialistic than we usually think: chemical reactions can induce thoughts, but thoughts can also induce chemical reactions.

Even so, the use of drugs for spiritual purposes feels cheap and

---

*Huxley suggests that the reason there aren't nearly as many mystics and visionaries walking around today, as compared to the Middle Ages, is the improvement in nutrition. Vitamin deficiencies wreak havoc on brain function and probably explain a large portion of visionary experiences in the past.

false. Perhaps it is our work ethic that is offended—you know, no pain, no gain. Or maybe it is the provenance of the chemicals that troubles us, the fact that they come *from outside*. Especially in the Judeo-Christian West, we tend to define ourselves by the distance we've put between ourselves and nature, and we jealousy guard the borders between matter and spirit as proof of our ties to the angels. The notion that spirit might turn out in some sense to *be* matter (and plant matter, no less!) is a threat to our sense of separateness and godliness. Spiritual knowledge comes from above or within, but surely not from plants. Christians have a name for someone who believes otherwise: pagan.

∷

Two stories stand behind the taboos that people in the West have placed on cannabis at various times in its history. Each reflects our anxieties about this remarkable plant, about what its Dionysian power might do to us if it is not resisted or brought under control.

The first, brought back from the Orient by Marco Polo (among others), is the story of the Assassins—or rather, a corruption of the story of the Assassins, which may or may not be apocryphal to begin with. The time is the eleventh century, when a vicious sect called the Assassins, under the absolute control of Hassan ibn al Sabbah (aka "the Old Man of the Mountain") is terrorizing Persia, robbing and murdering with brutal abandon. Hassan's marauders will do anything he tells them to, no questions asked; they have lost their fear of death. How does Hassan secure this perfect loyalty? By treating his men to a foretaste of the eternal paradise that will be theirs should they die in his service.

Hassan would begin his initiation of new recruits by giving them so much hashish that they passed out. Hours later the men would awaken to find themselves in the midst of a most beautiful palace garden, laid with sumptuous delicacies and staffed with gorgeous maidens to gratify their every desire. Scattered through this paradise, lying on the ground in pools of blood, are severed heads—actually actors buried to their necks. The heads speak, telling the men of the afterlife and what they will have to do if they hope ever to return to this paradise.

The story was corrupted by the time Marco Polo retold it, so that the hashish was now directly responsible for the violence of the Assassins. (The word itself is a corruption of "hashish.") By erasing the Assassins' fear of death, the story suggested, hashish freed them to commit the most daring and merciless crimes. The tale became a staple of orientalism and, later, of the campaign to criminalize marijuana in America in the 1930s. Harry J. Anslinger, the first director of the Federal Bureau of Narcotics and the man most responsible for marijuana prohibition, mentioned the Assassins at every opportunity. He skillfully used this metanarrative—publicizing every contemporary crime story he could cut to its lurid pattern—to transform a little-known drug of indolence into one of violence, a social menace. Even after Anslinger's "reefer madness" had subsided, the moral of the tale of the Assassins continued to trail cannabis—the notion that, by severing the link between acts and their consequences, marijuana unleashes human inhibitions, thereby endangering Western civilization.

The second story is simply this: In 1484, Pope Innocent VIII issued a papal condemnation of witchcraft in which he specifically condemned the use of cannabis as an "antisacrament" in satanic worship. The black mass celebrated by medieval witches and sorcerers presented a mocking mirror image of the Catholic Eucharist, and in it cannabis traditionally took the place of wine—serving as a pagan sacrament in a counterculture that sought to undermine the establishment church.

The fact that witches and sorcerers were the first Europeans to exploit the psychoactive properties of cannabis probably sealed its fate in the West as a drug identified with feared outsiders and cultures conceived in opposition: pagans, Africans, hippies. The two stories fed each other and in turn the plant's power: people who smoked cannabis were Other, and the cannabis they smoked threatened to let their Otherness loose in the land.

::||::

Witches the Church simply burned at the stake, but something more interesting happened to the witches' magic plants. The plants were too precious to banish from human society, so in the decades after Pope Innocent's fiat against witchcraft, cannabis, opium, belladonna, and the rest were simply transferred from the realm of sorcery to medicine, thanks largely to the work of a sixteenth-century Swiss alchemist and physician named Paracelsus. Sometimes called the "Father of Medicine," Paracelsus established a legitimate pharmacology largely on the basis of the ingredients found in flying ointments. (Among his many accomplishments was the invention of laudanum, the tincture of opium that was perhaps the most important drug in the pharmacopoeia until the twentieth century.) Paracelsus often said that he had learned everything he knew about medicine from the sorceresses. Working under the rational sign of Apollo, he domesticated their forbidden Dionysian knowledge, turning the pagan potions into healing tinctures, bottling the magic plants and calling them medicines.

Paracelsus's grand project, which arguably is still going on today,* represents one of the many ways the Judeo-Christian tra-

---

*Most recently, as the medical value of marijuana has been rediscovered, medicine has been searching for ways to "pharmaceuticalize" the plant—find a way to harness its easily accessible benefits in a patch or inhaler that doctors can prescribe, corporations patent, and governments regulate. Whenever possible,

dition has deployed its genius to absorb, or co-opt, the power of the pagan faith it set out to uproot. In much the same way that the new monotheism folded into its rituals the people's traditional pagan holidays and spectacles, it desperately needed to do something about their ancient devotion to magic plants. Indeed, the story of the forbidden fruit in Genesis suggests that nothing was more important.

The challenge these plants posed to monotheism was profound, for they threatened to divert people's gaze from the sky, where the new God resided, down to the natural world all around them. The magic plants were, and remain, a gravitational force pulling us back to Earth, to matter, away from the there and then of Christian salvation and back to the here and now. Indeed, what these plants do to time is perhaps the most dangerous thing about them—dangerous, that is, from the perspective of a civilization organized on the lines of Christianity and, more recently, capitalism.

Christianity and capitalism are both probably right to detest a plant like cannabis. Both faiths bid us to set our sights on the future; both reject the pleasures of the moment and the senses in favor of the expectation of a fulfillment yet to come—whether by earning salvation or by getting and spending. More even than most plant drugs, cannabis, by immersing us in the present and offering something like fulfillment here and now, short-circuits the metaphysics of desire on which Christianity and capitalism (and so much else in our civilization) depend.*

Paracelsus's lab-coated descendants have synthesized the active ingredients in plant drugs, allowing medicine to dispense with the plant itself—and any reminders of its pagan past.

*David Lenson draws a useful distinction between drugs of desire (cocaine, for example) and drugs of pleasure, such as cannabis. "Cocaine promises the

⠆⠆

What, then, *was* the knowledge that God wanted to keep from Adam and Eve in the Garden? Theologians will debate this question without end, but it seems to me the most important answer is hidden in plain sight. The *content* of the knowledge Adam and Eve could gain by tasting of the fruit does not matter nearly as much as its form—that is, the very fact that there was spiritual knowledge of *any* kind to be had from a tree: from nature. The new faith sought to break the human bond with magic nature, to disenchant the world of plants and animals by directing our attention to a single God in the sky. Yet Jehovah couldn't very well pretend the tree of knowledge didn't exist, not when generations of plant-worshiping pagans knew better. So the pagan tree is allowed to grow even in Eden, though ringed around now with a strong taboo. Yes, there *is* spiritual knowledge in nature, the new God is acknowledging, and its temptations are fierce, but I am fiercer still. Yield to it, and you will be punished.

So unfolds the drug war's first battle.

⠆⠆

I've removed most all of the temptations from my own garden, though not without regret or protest. Immersed this spring in research for this chapter, I was sorely tempted to plant one of the hybrid cannabis seeds I'd seen for sale in Amsterdam. I immediately

---

greatest pleasure ever known in just a minute more . . . But that future never comes." In this respect the cocaine experience is "a savage mimicry of consumer consciousness." With cannabis or the psychedelics, on the other hand, "pleasure can come from natural beauty, domestic tasks, friends and relatives, conversation, or any number of objects that do not need to be purchased."

thought better of it, however. So I planted lots of opium poppies instead. I hasten to add that I've no plans to do anything with my poppies except admire them—first their fleeting tissue-paper blooms, then their swelling blue-green seedpods, fat with milky alkaloid. (Unless, of course, simply walking among the poppies is enough to have an effect, as it was for Dorothy in Oz.) Unscored and so at least arguably innocent, these poppies are my stand-ins for the cannabis I cannot plant. Whenever I look at their dreamy petals, I'll be reminded of the powers this garden has abjured in order to stay on the safe side of the law.

So I make do with this bowdlerized garden, this densely planted plot of acceptable pleasures—good things to eat, beautiful things to gaze upon—fenced around by heeded laws. If Dionysus is represented in this garden, and he surely is, it's mainly in the flower border. I would be the last person to make light of the power of a fragrant rose to raise one's spirits, summon memories, even, in some not merely metaphorical sense, to intoxicate.

The garden is a place of many sacraments, an arena—at once as common as any room and as special as a church—where we can go not just to witness but to enact in a ritual way our abiding ties to the natural world. Abiding, yet by now badly attenuated, for civilization seems bent on breaking or at least forgetting our connections to the earth. But in the garden the old bonds are preserved, and not merely as symbols. So we eat from the vegetable patch, and, if we're paying attention, we're recalled to our dependence on the sun and the rain and the everyday leaf-by-leaf alchemy we call photosynthesis. Likewise, the poultice of comfrey leaves that lifts a wasp's sting from our skin returns us to a quasi-magic world of healing plants from which modern medicine would cast us out. Such sacraments are so benign that few of us have any trouble embracing them, even if they do sound a faintly pagan note. I'd guess that's because we're generally willing to be reminded that our bodies, at least, remain linked in such ways to the world of plants and animals, to nature's cycles.

But what about our minds? Here we're not so sure anymore. To take a leaf or flower and use it to change our experience of consciousness suggests a very different sort of sacrament, one at odds with our loftier notions of self, not to mention civilized society. But I'm inclined to think that such a sacrament may on occasion be worthwhile just the same, if only as a check on our hubris. Plants with the power to revise our thoughts and perceptions, to provoke metaphor and wonder, challenge the cherished Judeo-Christian belief that our conscious, thinking selves somehow stand apart from nature, have achieved a kind of transcendence.

Just what happens to this flattering self-portrait if we discover that transcendence itself owes to molecules that flow through our brains and at the same time through the plants in the garden? If some of the brightest fruits of human culture are in fact rooted deeply in this black earth, with the plants and fungi? Is matter, then, still as mute as we've come to think? Does it mean that spirit too is part of nature?

There may be no older idea in the world. Friedrich Nietzsche once described Dionysian intoxication as "nature overpowering mind"—nature having her way with us. The Greeks understood that this was not something to be undertaken lightly or too often. Intoxication was a carefully circumscribed ritual for them, never a way to live, because they understood that Dionysus can make angels of us or animals, it all depends. Even so, letting nature have her way with us now and again still seems like a useful thing to do, if only to bring our abstracted upward gaze back down to Earth for a time. What a reenchantment of the world that would be, to look around and see that the plants and the trees of knowledge grow in the garden still.

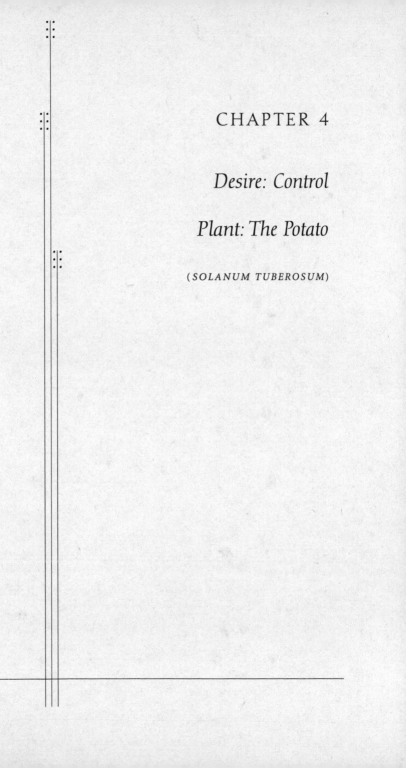

CHAPTER 4

Desire: Control

Plant: The Potato

(*SOLANUM TUBEROSUM*)

To my eye, there are few sights in nature quite as stirring as fresh rows of vegetable seedlings rising like a green city on the spring ground. I love the on-off digital rhythm of new green plant and black turned loam, the geometrical ordering of bounded earth that is the vegetable garden in May—before the plagues, before the rampancy, before the daunting complexities of summer. The sublimities of wilderness have their place, okay, and their legions of American poets, God knows, but I want to speak a word here for the satisfactions of the ordered earth. I'd call it the Agricultural Sublime if that didn't sound too much like an oxymoron.

Which it probably is. The experience of the sublime is all about nature having her way with us, about the sensation of awe before her power—about feeling small. What I'm talking about is the opposite, and admittedly more dubious, satisfaction of having our way with nature: the pleasure of beholding the reflection of our labor and intelligence in the land. In the same way that Niagara or Everest stirs the first impulse, the farmer's methodical rows stitching the hills, or the allées of pollarded trees ordering a garden like Versailles, excite the second, filling us with a sense of *our* power.

These days the sublime is mostly a kind of vacation, in both a literal and a moral sense. After all, who has a bad word to say about wilderness anymore? By comparison, this other impulse, the desire to exert our control over nature's wildness, bristles with ambiguity. We're unsure about our power in nature, its legitimacy, and its reality, and rightly so. Perhaps more than most, the farmer or the gardener understands that his control is always something of a fiction, depending as it does on luck and weather and much else that is beyond his control. It is only the suspension of disbelief that allows him to plant again every spring, to wade out in the season's uncertainties. Before long the pests will come, the storms and droughts and blights, as if to remind him just how imperfect the human power implied by those pristine rows really is.

In 1999 a freak December windstorm, more powerful than any Europeans could remember, laid waste to many of André Lenôtre's centuries-old plantings at Versailles, crumpling in a matter of seconds that garden's perfect geometries—perhaps as potent an image of human mastery as we have. When I saw the pictures of the wrecked allées, the straight lines scrabbled, the painterly perspectives ruined, it occurred to me that a less emphatically ordered garden would have been better able to withstand the storm's fury and repair itself afterward. So what are we to make of such a disaster? It all depends: on whether one regards that particular storm as a straightforward proof of our hubris and nature's infinitely superior power or, as some scientists now do, as an effect of global warming, which is adding to the atmosphere's instability. In that view, the storm is as much a human artifact as the order of trees it shattered, one manifestation of human power pulling the rug out from under another.

Ironies of this kind are second nature to the gardener, who eventually learns that every advance in his control of the garden is also an invitation to a new disorder. Wilderness might be reducible, acre by acre, but wildness is something else again. So the freshly hoed earth invites a new crop of weeds, the potent new pesticide engenders resistance in pests, and every new step in the direction of simplification—toward monoculture, say, or genetically identical plants—leads to unimagined new complexities.

Yet these simplifications are undeniably powerful: often as not, they "work"—get us what we want from nature. Agriculture is, by its very nature, brutally reductive, simplifying nature's incomprehensible complexity to something humanly manageable; it begins, after all, with the simple act of banishing all but a tiny handful of chosen species. Planting these in intelligible rows not only flatters our sense of order, it makes good sense too: weeding and harvesting become that much simpler. And though nature herself never plants in rows—or parterres or allées—she doesn't necessarily begrudge us when we do.

In fact, lots of new things happen in the garden, novelties unknown in nature before our attempts to exert control: edible potatoes (the wild ones are too bitter and toxic to eat), doubled tulips, sinsemilla, nectarines, to name a few. In every case nature supplied the necessary genes or mutations, but without the garden and the gardener to make a space for these novelties, they would never have seen the light of day.

For nature as much as for people, the garden has always been a place to experiment, to try out new hybrids and mutations. Species that never cross in the wild will freely hybridize on land cleared by people. That's because a novel hybrid has a hard time finding a purchase in the tight weave of an established meadow or forest ecosystem; every possible niche is apt to be already filled. But a garden—or a roadside or a dump heap—is by comparison an "open" habitat in which a new hybrid has a much better shot, and if it happens to catch our fancy, to gratify a human desire, it stands to make its way in the world. One theory of the origins of agriculture holds that domesticated plants first emerged on dump heaps, where the discarded seeds of the wild plants that people gathered and ate—already unconsciously selected for sweetness or size or power—took root, flourished, and eventually hybridized. In time people gave the best of these hybrids a place in the garden, and there, together, the people and the plants embarked on a series of experiments in coevolution that would change them both forever.

⁝⁝

The garden is still a site for experiment, a good place to try out new plants and techniques without having to bet the farm. Many of the methods employed by organic farmers today were first discovered in the garden. Attempted on the scale of a whole farm, the next New Thing is an expensive and risky proposition, which is why farmers have always been a conservative breed, notoriously slow to change. But for a gardener like me, with relatively little at stake, it's no big deal to try out a new variety of potato or method of pest control, and every season I do.

Admittedly, my experiments in the garden are unscientific and far from foolproof or conclusive. Is it the new neem tree oil I sprayed on the potatoes that's controlling the beetles so well this year, or the fact I planted a pair of tomatillos nearby, the leaves of which the beetles seem to prefer to potatoes? (My scapegoats, I call them.) Ideally, I'd control for every variable but one, but that's hard to do in a garden, a place that, like the rest of nature, seems to consist of nothing *but* variables. "Everything affecting everything else" is not a bad description of what happens in a garden or, for that matter, in any ecosystem.

In spite of these complexities, it is only by trial and error that my garden ever improves, so I continue to experiment. Recently I planted something new—something very new, as a matter of fact—and embarked on my most ambitious experiment to date. I planted a potato called "NewLeaf" that has been genetically engineered (by the Monsanto corporation) to produce its own insecticide. This it does in every cell of every leaf, stem, flower, root, and—this is the unsettling part—every spud.

The scourge of potatoes has always been the Colorado potato beetle, a handsome, voracious insect that can pick a plant clean of its leaves virtually overnight, starving the tubers in the process. Supposedly, any Colorado potato beetle that takes so much as a nibble of a NewLeaf leaf is doomed, its digestive tract pulped, in effect, by the bacterial toxin manufactured in every part of these plants.

I wasn't at all sure I really *wanted* the NewLeaf potatoes I'd be digging at the end of the season. In this respect my experiment in growing them was very different from anything else I've ever done in my garden—whether growing apples or tulips or even pot. All of those I'd planted because I really wanted what the plants promised. What I wanted here was to gratify not so much a desire as a curiosity: Do they work? Are these genetically modified potatoes a good idea, either to plant or to eat? If not mine, then whose desire *do* they gratify? And finally, what might they have to tell us about the future of the relationship between plants and people? To answer these questions, or at least begin to, would take more than the tools of the gardener (or the eater); I'd need as well the tools of the journalist, without which I couldn't hope to enter the world from which these potatoes had come. So you could say there was something fundamentally artificial about my experiment in growing NewLeaf potatoes. But then, artificiality seems very much to the point.

::|::

Certainly my NewLeafs are aptly named. They're part of a new class of crop plant that is transforming the long, complex, and by now largely invisible food chain that links every one of us to the land. By the time I conducted my experiment, more than fifty million acres of American farmland had already been planted to genetically modified crops, most of it corn, soybeans, cotton, and potatoes that have been engineered either to produce their own pesticide or to withstand herbicides. The not-so-distant future will, we're told, bring us potatoes genetically modified to absorb less fat when fried, corn that can withstand drought, lawns that don't ever have to be mowed, "golden rice" rich in Vitamin A, bananas and potatoes that deliver vaccines, tomatoes enhanced with flounder genes (to withstand frost), and cotton that grows in every color of the rainbow.

It's probably not too much to say that this new technology represents the biggest change in the terms of our relationship with plants since people first learned how to cross one plant with another. With genetic engineering, human control of nature is taking a giant step forward. The kind of reordering of nature represented by the rows in a farmer's field can now take place at a whole new level: within the genome of the plants themselves. Truly, we have stepped out onto new ground.

Or have we?

Just how novel these plants really are is in fact one of the biggest questions about them, and the companies that have developed them give contradictory answers. The industry simultaneously depicts these plants as the linchpins of a biological revolution— part of a "paradigm shift" that will make agriculture more sustainable and feed the world—and, oddly enough, as the same old spuds, corn, and soybeans, at least so far as those of us at the eating end of the food chain should be concerned. The new plants are novel enough to be patented, yet not so novel as to warrant a label telling us what it is we're eating. It would seem they are chimeras: "revolutionary" in the patent office and on the farm, "nothing new" in the supermarket and the environment.

By planting my own crop of NewLeafs, I was hoping to figure out which version of reality to believe, whether these were indeed the same old spuds or something sufficiently novel (in nature, in the diet) to warrant caution and hard questions. As soon as you start looking into the subject, you find that there are many questions about genetically modified plants that, fifty million acres later, remain unanswered and, more remarkable still, unasked— enough to make me think mine might not be the only experiment going on.

∷

May 2. Here at the planter's end of the food chain, where I began my experiment after Monsanto agreed to let me test-drive its NewLeafs, things certainly look new and different. After digging two shallow trenches in my vegetable garden and lining them with compost, I untied the purple mesh bag of seed potatoes Monsanto had sent and opened the grower's guide tied around its neck. Potatoes, you will recall from kindergarten experiments, are grown not from actual seeds but from the eyes of other potatoes, and the dusty, stone-colored chunks of tuber I carefully laid at the bottom of the trench looked much like any other. Yet the grower's guide that comes with them put me in mind not so much of planting vegetables as booting up a new software release.

By "opening and using this product," the card informed me, I was now "licensed" to grow these potatoes, but only for a single generation; the crop I would water and tend and harvest was mine, yet also not mine. That is, the potatoes I would dig come September would be mine to eat or sell, but their genes would remain the intellectual property of Monsanto, protected under several U.S. patents, including 5,196,525; 5,164,316; 5,322,938; and 5,352,605. Were I to save even one of these spuds to plant next year—something I've routinely done with my potatoes in the past—I would be breaking federal law. (I had to wonder, what would be the legal status of any "volunteers"—those plants that, with no prompting from the gardener, sprout each spring from tubers overlooked during the previous harvest?) The small print on the label also brought the disconcerting news that my potato plants were *themselves* registered as a pesticide with the Environmental Protection Administration (U.S. EPA Reg. No. 524-474).

If proof were needed that the food chain that begins with seeds and ends on our dinner plates is in the midst of revolutionary change, the small print that accompanied my NewLeafs will do. That food chain has been unrivaled for its productivity: on average, an American farmer today grows enough food each year to feed a hundred people. Yet that achievement—that power over nature—has come at a price. The modern industrial farmer cannot grow that much food without large quantities of chemical fertilizers, pesticides, machinery, and fuel. This expensive set of "inputs," as they're called, saddles the farmer with debt, jeopardizes his health, erodes his soil and ruins its fertility, pollutes the groundwater, and compromises the safety of the food we eat. Thus the gain in the farmer's power has been trailed by a host of new vulnerabilities.

All this I'd heard before, of course, but always from environmentalists or organic farmers. What is new is to hear the same critique from industrial farmers, government officials, and the agribusiness companies that sold farmers on all those expensive inputs in the first place. Taking a page from Wendell Berry, of all people, Monsanto declared in a recent annual report that "current agricultural technology is unsustainable."

What is to rescue the American food chain is a new kind of plant. Genetic engineering promises to replace expensive and toxic chemicals with expensive but apparently benign genetic information: crops that, like my NewLeafs, can protect themselves from insects and diseases without the help of pesticides. In the case of the NewLeaf, a gene borrowed from one strain of a common bacterium found in the soil—*Bacillus thuringiensis,* or "Bt" for short—gives the potato plant's cells the information they need to manufacture a toxin lethal to the Colorado potato beetle. This gene is now Monsanto's intellectual property. With genetic engineering, agriculture has entered the information age, and Monsanto's aim, it would appear, is to become its Microsoft, supplying the proprietary "operating systems"—the metaphor is theirs—to run this new generation of plants.

The metaphors we use to describe the natural world strongly influence the way we approach it, the style and extent of our attempts at control. It makes all the difference in (and to) the world if one conceives of a farm as a factory or a forest as a farm. Now we're about to find out what happens when people begin approaching the genes of our food plants as software.

∷

The Andes, 1532. The patented potatoes I was planting are descended from wild ancestors growing on the Andean altiplano, the potato's "center of diversity." It was here that *Solanum tuberosum* was first domesticated more than seven thousand years ago by ancestors of the Incas. Actually, some of the potatoes in my garden are closely related to those ancient potatoes. Among the half-dozen or so different varieties I grow are a couple of ancient heirlooms, including the Peruvian blue potato. This starchy spud is about the size of a golf ball; when you slice it through the middle the flesh looks as though it has been tie-dyed the most gorgeous shade of blue.

My blue potato is part of the cornucopia of potatoes developed by the Incas along with their ancestors and descendants. In addition to the blue potato, the Incas grew reds, pinks, yellows, and oranges; all manner of skinnies and fatties, smooth-skinneds and russets, short-season spuds and long, drought-tolerant and water-loving, sweet tubers and bitter ones (good for forage), starchy potatoes and others almost buttery in texture—some three thousand different spuds in all. This extravagant flowering of potato diversity owes partly to the Incas' desire for variety, partly to their flair for experimentation, and partly to the intricacy of their agriculture, the most sophisticated in the world at the time of the Spanish conquest. While I was waiting for my potatoes to come up that May, I began reading about theirs (and then those of the Irish), hoping to get a clearer picture of the relationship between people and potatoes, and how that relationship had changed both the plant and ourselves.

The Incas figured out how to grow impressive yields of potatoes under the most inauspicious conditions, developing an approach that is still in use in parts of the Andes today. A more or less vertical habitat presents special challenges to both plants and their cultivators, because the microclimate changes dramatically with every change in altitude or orientation to the sun and wind. A potato that thrives on one side of a ridge at one altitude will languish in another plot only a few steps away. No monoculture could succeed under such circumstances, so the Incas developed a method of farming that is monoculture's exact opposite. Instead of betting the farm on a single cultivar, the Andean farmer, then as now, made a great many bets, at least one for every ecological niche. Instead of attempting, as most farmers do, to change the environment to suit a single optimal spud—the Russet Burbank, say—the Incas developed a different spud for every environment.

To Western eyes, the resulting farms look patchy and chaotic; the plots are discontinuous (a little of this growing here, a little of that over there), offering none of the familiar, Apollonian satisfactions of an explicitly ordered landscape. Yet the Andean potato farm represented an intricate ordering of nature that, unlike Versailles in 1999, say, or Ireland in 1845, can withstand virtually anything nature is apt to throw at it.

Since the margins and hedgerows of the Andean farm were, and still are, populated by weedy wild potatoes, the farmer's cultivated varieties have regularly crossed with their wild relatives, in the process refreshing the gene pool and producing new hybrids. Whenever one of these new potatoes proves its worth—surviving a drought or storm, say, or winning praise at the dinner table—it is promoted from the margins to the fields and, in time, to the neighbors' fields as well. Artificial selection is thus a continual local process, each new potato the product of an ongoing back-and-forth between the land and its cultivators, mediated by the universe of all possible potatoes: the species' genome.

The genetic diversity cultivated by the Incas and their descendants is an extraordinary cultural achievement and a gift of incalculable value to the rest of the world. A free and unencumbered gift, one might add, quite unlike my patented and trademarked NewLeafs. "Intellectual property" is a recent, Western concept that means nothing to a Peruvian farmer, then or now.* Of course, Francisco Pizarro was looking for neither plants nor intellectual property when he conquered the Incas; he had eyes only for gold. None of the conquistadores could have imagined it, but the funny-looking tubers they encountered high in the Andes would prove to be the single most important treasure they would bring back from the New World.

∷

May 15. After several days of drenching rain, the sun appeared this week, and so did my NewLeafs: a dozen deep green shoots pushed up out of the soil and commenced to grow—faster and more robustly than any of my other potatoes. Apart from their vigor, though, my NewLeafs looked perfectly normal—they certainly didn't beep or glow, as a few visitors to my garden jokingly inquired. (Not that the glowing notion is so far-fetched: I've read that plant breeders have developed a luminescent tobacco plant by inserting a gene from a firefly. I've yet to read *why* they would do this, except perhaps to prove it could be done: a demonstration of power.) Yet as I watched my NewLeafs multiply their lustrous, dark green leaves those first few days, eagerly awaiting the arrival of the first unwitting beetle, I couldn't help thinking of them as existentially different from the rest of my plants.

---

*In fact, "intellectual property" has been defined under recent trade agreements in such a way as to specifically exclude any innovations that are not the private, marketable property of an individual or corporation. Thus a corporation's new potato qualifies as intellectual property, but not a tribe's.

All domesticated plants are in some sense artificial, living archives of both cultural and natural information that people have helped to "design." Any given type of potato reflects the human desires that have been bred into it. One that's been selected to yield long, handsome french fries or unblemished, round potato chips is the expression of a national food chain and a culture that likes its potatoes highly processed. At the same time, some of the more delicate European fingerlings growing beside my NewLeafs imply an economy of small-market growers and a cultural taste for eating potatoes fresh—for none of these varieties can endure much travel or time in storage. I'm not sure exactly what cultural values to ascribe to my Peruvian blues; perhaps nothing more than a craving for variety among a people who ate potatoes morning, noon, and night.

"Tell me what you eat," Anthelme Brillat-Savarin famously claimed, and "I will tell you what you are." The qualities of a potato—as of any domesticated plant or animal—are a fair reflection of the values of the people who grow and eat it. Yet all these qualities already existed in the potato, somewhere within the universe of genetic possibilities presented by the species *Solanum tuberosum*. And though that universe may be vast, it is not infinite. Since unrelated species in nature cannot be crossed, the breeder's art has always run up against a natural limit of what a potato is willing, or able, to do—that species' essential identity. Nature has always exercised a kind of veto over what culture can do with a potato.

Until now. The NewLeaf is the first potato to override that veto. Monsanto likes to depict genetic engineering as just one more chapter in the ancient history of human modifications of nature, a story going back to the discovery of fermentation. The company defines the word *biotechnology* so broadly as to take in the brewing of beer, cheese making, and selective breeding: all are "technologies" that involve the manipulation of life-forms.

Yet this new biotechnology has overthrown the old rules governing the relationship of nature and culture in a plant. Domestication has never been a simple one-way process in which our species has controlled others; other species participate only so far as their interests are served, and many plants (such as the oak) simply sit the whole game out. That game is the one Darwin called "artificial selection," and its rules have never been any different from the rules that govern natural selection. The plant in its wildness proposes new qualities, and then man (or, in the case of natural selection, nature) selects which of those qualities will survive and prosper. But about one rule Darwin was emphatic; as he wrote in *The Origin of Species,* "Man does not actually produce variability."

Now he does. For the first time, breeders can bring qualities at will from anywhere in nature into the genome of a plant: from fireflies (the quality of luminescence), from flounders (frost tolerance), from viruses (disease resistance), and, in the case of my potatoes, from the soil bacterium known as *Bacillus thuringiensis.* Never in a million years of natural or artificial selection would these species have proposed those qualities. "Modification by descent" has been replaced by . . . something else.

Now, it is true that genes occasionally move between species; the genome of many species appears to be somewhat more fluid than scientists used to think. Yet for reasons we don't completely understand, distinct species do exist in nature, and they exhibit a certain genetic integrity—sex between them, when it does occur, doesn't produce fertile offspring. Nature presumably has some reason for erecting these walls, even if they are permeable on occasion. Perhaps, as some biologists believe, the purpose of keeping species separate is to put barriers in the path of pathogens, to contain their damage so that a single germ can't wipe out life on Earth at a stroke.

The deliberate introduction into a plant of genes transported not only across species but across whole phyla means that the wall of that plant's essential identity—its irreducible wildness, you might say—has been breached, not by a virus, as sometimes happens in nature, but by humans wielding powerful new tools.

For the first time the genome itself is being domesticated— brought under the roof of human culture. This made the potato I was growing slightly different from the other plants in this book, all of which had been both the subjects and the objects of domestication. While the other plants coevolved in a kind of conversational give-and-take with people, the NewLeaf potato has really only taken, only listened. It may or may not profit from the gift of its new genes; we can't yet say. What we can say, though, is that this potato is not the hero of its own story in quite the same way the apple has been. It didn't come up with this Bt scheme all on its evolutionary own. No, the heroes of the NewLeaf story are scientists working for Monsanto. Certainly the scientists in the lab coats have something in common with the fellow in the coffee sack: both work, or worked, at disseminating plant genes around the world. Yet although Johnny Appleseed and the brewers of beer and makers of cheese, the high-tech pot growers and all the other "biotechnologists" manipulated, selected, forced, cloned, and otherwise altered the species they worked with, the species themselves never lost their evolutionary say in the matter—never became solely the objects of our desires. Now the once irreducible wildness of these plants has been . . . reduced. Whether this is a good or bad thing for the plants (or for us), it is unquestionably a *new* thing.

What is perhaps most striking about the NewLeafs coming up in my garden is the added human intelligence that the insertion of the *Bacillus thuringiensis* gene represents. In the past that intelligence resided outside the plant, in the minds of the organic farmers and gardeners (myself included) who used Bt, commonly in the form of a spray, to manipulate the ecological relationship between certain insects and a certain bacterium in order to foil those insects. The irony about the new Bt crops (a similar gene has been inserted into corn plants) is that the cultural information they encode happens to be knowledge that's always resided in the heads of the very sorts of people—that is, organic growers—who most distrust high technology. Most of the other biotech crops—such as the ones Monsanto has engineered to withstand Roundup, the company's patented herbicide—encode a very different, more industrial sort of intelligence.

One way to look at genetic engineering is that it allows a larger portion of human culture and intelligence to be incorporated into the plants themselves. From this perspective, my NewLeafs are just plain smarter than the rest of my potatoes. The others will depend on my knowledge and experience when the Colorado potato beetles strike. The NewLeafs, already knowing what I know about bugs and Bt, will take care of themselves. So while my genetically engineered plants might at first seem like alien beings, that's not quite right; they're more like us than other plants because there's more of us in them.

:‖:

Ireland, 1588. Like an alien species introduced into an established ecosystem, the potato had trouble finding a foothold when it first arrived in Europe toward the end of the sixteenth century, probably as an afterthought in the hold of a Spanish ship. The problem was not with the European soil or climate, which would prove very much to the potato's liking (in the north anyway), but with the European mind. Even after people recognized that this peculiar new plant could produce more food on less land than any other crop, most of European culture remained inhospitable to the potato. Why? Europeans hadn't eaten tubers before; the potato was a member of the nightshade family (along with the equally disreputable tomato); potatoes were thought to cause leprosy and immorality; potatoes were mentioned nowhere in the Bible; potatoes came from America, where they were the staple of an uncivilized and conquered race. The justifications given for refusing to eat potatoes were many and diverse, but in the end most of them came down to this: the new plant—and in this respect it was quite unlike my NewLeaf—seemed to contain in its being too little of human culture and rather too much unreconstructed nature.

Oh, but what about Ireland? Ireland was the exception that proved the rule—indeed, the exception that largely wrote the rule, since that country's extraordinary relationship to the potato consolidated its dubious identity in the English mind. Ireland embraced the potato very soon after its introduction, a fateful event sometimes credited to Sir Walter Raleigh, sometimes to the shipwreck of a Spanish galleon off the Irish coast in 1588. As it happened, the cultural, political, and biological environment of Ireland could not have better suited the new plant. Cereal grains grow poorly on the island (wheat hardly at all), and, in the seventeenth century, Cromwell's Roundheads seized what little arable land there was for English landowners, forcing the Irish peasantry to eke out a subsistence from soil so rain-soaked and stingy that virtually nothing would grow in it. The potato, miraculously, would, managing to extract prodigious amounts of food from the very land the colonial English had given up on. And so, by the end of the seventeenth century, the plant had made a beachhead in the Old World; within two centuries it would overrun northern Europe, in the process substantially remaking its new habitat.

The Irish discovered that a few acres of marginal land could produce enough potatoes to feed a large family and its livestock. The Irish also found they could grow these potatoes with a bare minimum of labor or tools, in something called a "lazy bed." The spuds were simply laid out in a rectangle on the ground; then, with a spade, the farmer would dig a drainage trench on either side of his potato bed, covering the tubers with whatever soil, sod, or peat came out of the trench. No plowed earth, no rows, and certainly no Agricultural Sublime—a damnable defect in English eyes. Potato growing looked nothing like agriculture, provided none of the Apollonian satisfactions of an orderly field of grain, no martial ranks of golden wheat ripening in the sun. Wheat pointed up, to the sun and civilization; the potato pointed down. Potatoes were chthonic, forming their undifferentiated brown tubers unseen beneath the ground, throwing a slovenly flop of vines above.

The Irish were too hungry to worry about agricultural aesthetics. The potato might not have presented a picture of order or control in the field, yet it gave the Irish a welcome measure of control over their lives. Now they could feed themselves off the economic grid ruled by the English and not have to worry so much about the price of bread or the going wage. For the Irish had discovered that a diet of potatoes supplemented with cow's milk was nutritionally complete. In addition to energy in the form of carbohydrates, potatoes supplied considerable amounts of protein and vitamins B and C (the spud would eventually put an end to scurvy in Europe); all that was missing was vitamin A, and that a bit of milk could make up. (So it turns out that mashed potatoes are not only the ultimate comfort food but all a body really needs.) And as easy as they were to grow, potatoes were even easier to prepare: dig, heat—by either boiling them in a pot or simply dropping them into a fire—and eat.

Eventually the potato's undeniable advantages over grain would convert all of northern Europe, but outside Ireland the process was never anything less than a struggle. In Germany, Frederick the Great had to force peasants to plant potatoes; so did Catherine the Great in Russia. Louis XVI took a subtler tack, reasoning that if he could just lend the humble spud a measure of royal prestige, peasants would experiment with it and discover its virtues. So Marie Antoinette took to wearing potato flowers in her hair, and Louis hatched an ingenious promotional scheme. He ordered a field of potatoes planted on the royal grounds and then posted his most elite guard to protect the crop during the day. He sent the guards home at midnight, however, and in due course the local peasants, suddenly convinced of the crop's value, made off in the night with the royal tubers.

In time, all three nations would grow powerful on potatoes, which put an end to malnutrition and periodic famine in northern Europe and allowed the land to support a much larger population than it ever could have planted in grain. Since fewer hands were needed to farm it, the potato also allowed the countryside to feed northern Europe's growing and industrializing cities. Europe's center of political gravity had always been anchored firmly in the hot, sunny south, where wheat grew reliably; without the potato, the balance of European power might never have tilted north.

The last redoubt of antipotato prejudice was in England, and there it was not confined to a hidebound or superstitious peasantry. Well into the nineteenth century, a significant portion of elite opinion in London regarded the potato as nothing more or less than a threat to civilization. Proof? All one had to do was point in the direction of Ireland.

∷

England, 1794. The wheat harvest in the British Isles failed in 1794, sending the price of white bread beyond the reach of England's poor. Food riots broke out, and with them a great debate over the potato that would rage, on and off, for half a century. (The potato debate is recounted in Redcliffe Salaman's definitive 1949 volume, *The History and Social Influence of the Potato,* and its rhetoric is brilliantly dissected in "The Potato in the Materialist Imagination," an essay by the literary critic Catherine Gallagher.) Engaging the energies of the country's leading journalists, agronomists, and political economists, the potato debate brought to the surface predictable English anxieties about class conflict and the "Irish problem." But it also threw into sharp relief people's deepest feelings about their food plants and the ways they root us, for better and worse, in nature. Do we control these plants? Or do they control us?

The debate was kicked off by the potato's advocates, who argued that introducing a second staple would be a boon to England, a way to feed the poor when bread was dear and keep wages—which tended to track the price of bread—from rising. Arthur Young, a respected agronomist, had traveled to Ireland and returned convinced that the potato was "a root of plenty" that could protect England's poor from hunger and give farmers more control over their circumstances at a time when the enclosure movement was undermining their traditional way of life.

The radical journalist William Cobbett also traveled to Ireland, yet he returned with a very different picture of the potato eaters. Whereas Young had seen self-reliance in the Irishman's potato patch, Cobbett saw only abject subsistence and dependence. Cobbett argued that while it was true that the potato fed the Irish, it also impoverished them, by driving up the country's population—from three million to eight million in less than a century—and driving down its wages. The prolific potato allowed young Irishmen to marry earlier and support a larger family; as the labor supply increased, wages fell. The bounty of the potato was its curse.

In his articles Cobbett depicted "this damned root" as a kind of gravitational force, pulling the Irishman out of civilization and back down into the earth, gradually muddying the distinctions between man and beast, even man and root. This is how he described the potato eater's mud hut: "no windows at all; . . . the floor nothing but the bare earth; no chimney, but a hole at one end . . . surrounded by a few stones." In Cobbett's grim imagery, the Irish had themselves moved underground, joining their tubers in the mud. Once cooked, the potatoes "are taken up and turned into a great dish," Cobbett wrote. "The family squat round this basket and take out the potatoes with their hands; the pig stands and is helped by some one, and sometimes he eats out of the pot. He goes in and out and about the hole, like one of the family." The potato had single-handedly unraveled civilization, putting nature back in control of man.

"Bread root" was what the English sometimes called the potato, and the symbolic contrast between the two foods loomed large in the debate, never to the spud's advantage. Catherine Gallagher points out that the English usually depicted the potato as mere food, primitive, unreconstructed, and lacking in any cultural resonance. In time, that lack would itself become precisely the potato's cultural resonance: the potato came to signify the end of food being anything more than food—animal fuel. Bread, on the other hand, was as leavened with meaning as it was with air.

Like the potato, wheat begins in nature, but it is then transformed by culture. While the potato is simply thrown into a pot or fire, wheat must be harvested, threshed, milled, mixed, kneaded, shaped, baked, and then, in a final miracle of transubstantiation, the doughy lump of formless matter rises to become bread. This elaborate process, with its division of labor and suggestion of transcendence, symbolized civilization's mastery of raw nature. A mere food thus became the substance of human and even spiritual communion, for there was also the old identification of bread with the body of Christ. If the lumpish potato was base matter, bread in the Christian mind was its very opposite: antimatter, even spirit.

The political economists also weighed in on the potato debate, and though they framed their arguments in somewhat more scientific terms, their rhetoric too betrays deep anxieties about nature's threat to civilization's control. Malthusian logic started from the premise that people are driven by the desires for food and sex; only the threat of starvation keeps the population from exploding. The danger of the potato, Malthus believed, was that it removed the economic constraints that ordinarily kept the population in check. This in a nutshell was Ireland's problem: "the indolent and turbulent habits of the lower Irish can never be corrected while the potato system enables them to increase so much beyond the regular demand for labour."

In the same way that the potato exempts the potato eater from the civilizing processes of bread making, it also exempts him from the discipline of the economy. Political economists like Adam Smith and David Ricardo regarded the market as a sensitive mechanism for adjusting the size of the population to the demand for labor, and the price of bread was that mechanism's regulator. When the price of wheat rose, people had to curb both of their animal appetites and so produced fewer babies. The problem with "the potato system" is that, under it, the *Homo economicus* who adjusts his behavior to the algebra of need is replaced by a far less rational actor—*Homo appetitus,* as Gallagher calls him. If Economic Man operated under the coolly rational sign of Apollo, Appetite Man was in thrall to earthy, fecund, amoral Dionysus. Since the Irishman grew and ate his own potatoes, and since his potatoes (unlike wheat flour) could not easily be stored or traded, they never became commodities and were therefore, like him, subject to no authority but nature's own.

In the eyes of the political economists, capitalist exchange was a lot like baking, since it represented a way of civilizing anarchic nature—the anarchic nature, that is, of both plants and people. Without the discipline of commodity markets, man is thrown back on his instincts: unlimited food and sex leading inexorably to overpopulation and misery. David Ricardo was convinced that the potato was both the cause and symbol of this regression, this surrender of control to nature. As long as humans need to eat, we can never completely insulate ourselves from the vicissitudes of nature; the best we can do, Ricardo believed, was to rely on a staple that, like wheat, can be stored against storms and droughts and readily converted into money to buy other foods. The potato offered no such security. By refusing to transcend its own nature and become a commodity, the potato threatened, in Gallagher's words, to "wipe out the progress an advanced economy has made in liberating humankind from dependence on shifty nature."

About this much, at least, history would prove the political economists terribly correct. The control with which the potato appeared to have blessed the Irish would turn out to be a cruel illusion. Dependence on the potato had in fact made the Irish exquisitely vulnerable, not to the vicissitudes of the economy so much as to those of nature. This they would abruptly discover late in the summer of 1845, when *Phytophthora infestans* arrived in Europe, probably on a ship from America. Within weeks the spores of this savage fungus, borne on the wind, overspread the continent, dooming potatoes and potato eaters alike.

⁘

St. Louis, June 23. While my NewLeafs were bushing up nicely during a spell of hot early-summer weather, I traveled to Monsanto's headquarters in St. Louis, where the ancient, noble dream of control of nature is in full and extravagant flower. If the place to go to understand the relationship of people and potato was a mountainside farm in South America in 1532 or a lazy bed near Dublin in 1845, today it is just as surely a research greenhouse on a corporate campus outside St. Louis.

My NewLeafs are clones of clones of plants that were first engineered more than a decade ago in a long, low-slung brick building on the bank of the Missouri that would look like any other corporate complex if not for its stunning roofline. What appear from a distance to be shimmering crenellations of glass turn out to be the twenty-six greenhouses that crown the building in a dramatic sequence of triangular peaks. The first generation of genetically altered plants—of which the NewLeaf potato is one—has been grown under this roof, in these greenhouses, since 1984; especially in the early days of biotechnology, no one knew for sure if it was safe to grow these plants outdoors, in nature. Today this research and development facility is one of a small handful of such places—Monsanto has only two or three competitors in the world—where the world's crop plants are being redesigned.

Dave Starck, one of Monsanto's senior potato people, escorted me through the clean rooms where potatoes are genetically engineered. He explained that there are two ways of splicing foreign genes into a plant: by infecting it with agrobacterium, a pathogen whose modus operandi is to break into a plant cell's nucleus and replace its DNA with some of its own, or by shooting it with a gene gun. For reasons not yet understood, the agrobacterium method seems to work best on broadleaf species such as the potato, the gene gun better on grasses, such as corn and wheat.

The gene gun is a strangely high-low piece of technology, but the main thing you need to know about it is that the gun here is not a metaphor: a .22 shell is used to fire stainless-steel projectiles dipped in a DNA solution at a stem or leaf of the target plant. If all goes well, some of the DNA will pierce the wall of some of the cells' nuclei and elbow its way into the double helix: a bully breaking into a line dance. If the new DNA happens to land in the right place—and no one yet knows what, or where, that place is—the plant grown from that cell will express the new gene. *That's it?* That's it.

Apart from its slightly more debonair means of entry, the agrobacterium works in much the same way. In the clean rooms, where the air pressure is kept artificially high to prevent errant microbes from wandering in, technicians sit at lab benches before petri dishes in which fingernail-sized sections of potato stem have been placed in a clear nutrient jelly. Into this medium they squirt a solution of agrobacteria, which have already had their genes swapped with the ones Monsanto wants to insert (specific enzymes can be used to cut and paste precise sequences of DNA). In addition to the Bt gene being spliced, a "marker" gene is also included—typically this is a gene conferring resistance to a specific antibiotic. This way, the technicians can later flood the dish with the antibiotic to see which cells have taken up the new DNA; any that haven't simply die. The marker gene can also serve as a kind of DNA fingerprint, allowing Monsanto to identify its plants and their descendants long after they've left the lab. By performing a simple test on any potato leaf in my garden, a Monsanto agent can prove whether or not the plant is the company's intellectual property. I realized that, whatever else it is, genetic engineering is also a powerful technique for transforming plants into private property, by giving every one of them what amounts to its own Universal Product Code.

After several hours the surviving slips of potato stem begin to put down roots; a few days later, these plantlets are moved upstairs to the potato greenhouse on the roof. Here I met Glenda De-brecht, a cheerful staff horticulturist, who invited me to don latex gloves and help her transplant pinkie-sized plantlets from their petri dishes to small pots filled with customized soil. After the abstractions of the laboratory, I felt back on quasi-familiar ground, in a greenhouse handling actual plants.

The whole operation, from petri dish to transplant to greenhouse, is performed thousands of times, Glenda explained as we worked across a wheeled potting bench from each other, largely because there is so much uncertainty about the outcome, even after the DNA is accepted. If the new DNA winds up in the wrong place in the genome, for example, the new gene won't be expressed, or it will be expressed only poorly. In nature—that is, in sexual reproduction—genes move not one by one but in the company of associated genes that regulate their expression, turning them on and off. The transfer of genetic material is also much more orderly in sex, the process somehow ensuring that every gene ends up in its proper neighborhood and doesn't trip over other genes in the process, inadvertently affecting their function. "Genetic instability" is the catchall term used to describe the various unexpected effects that misplaced or unregulated foreign genes can have on their new environment. These can range from the subtle and invisible (a particular protein is over- or underexpressed in the new plant, say) to the manifestly outlandish: Glenda sees a great many freaky potato plants.

Starck told me that the gene transfer "takes" anywhere between 10 percent and 90 percent of the time—an eyebrow-raising statistic. For some unknown reason (genetic instability?), the process produces a great deal of variability, even though it begins with a single, known, cloned strain of potato. "So we grow out thousands of different plants," Glenda explained, "and then look for the best." The result is often a potato that is superior in ways the presence of the new gene can't explain. This would certainly explain the vigor of my NewLeafs.

I was struck by the uncertainty surrounding the process, how this technology is at the same time both astoundingly sophisticated yet still a shot in the genetic dark. Throw a bunch of DNA against the wall and see what sticks; do this enough times, and you're bound to get what you're looking for. Transplanting potatoes with Glenda also made me realize that it may be impossible ever to conclude once and for all that this technology is *intrinsically* sound or dangerous. For every new genetically engineered plant is a unique event in nature, bringing its own set of genetic contingencies. This means that the reliability or safety of one genetically modified plant doesn't necessarily guarantee the reliability or safety of the next.

"There's still a lot we don't understand about gene expression," Starck acknowledged. A great many factors, including the environment, influence whether, and to what extent, an introduced gene will do what it's supposed to do. In one early experiment, scientists succeeded in splicing a gene for redness into petunias. In the field everything went according to plan, until the temperature hit 90 degrees and an entire planting of red petunias suddenly and inexplicably turned white. Wouldn't this sort of thing—these Dionysian jokers rearing up in the ordered Apollonian fields—rattle one's faith in genetic determinism a *little*? It's obviously not quite as simple as putting a software program into a computer.

∴

July 1. When I got home from St. Louis, my potato crop was thriving. It was time to hill up the plants, so, with a hoe, I pulled the rich soil from the lips of the trenches down around the stems to protect the developing tubers from the light. I also dressed the plants with a few shovelfuls of old cow manure: potatoes seem to love the stuff. The best, sweetest potatoes I ever tasted were ones that, as a teenager, I helped a neighbor dig out of the pile of pure horse manure he'd planted them in. I sometimes think it must have been this dazzling example of alchemy that sold me—not just on potato growing but on gardening as a quasi-magical, quasi-sacramental thing to do.

My NewLeafs were big as shrubs now, and crowned with slender flower stalks. Potato flowers are actually quite pretty, at least by the standards of a vegetable: five-petaled lavender stars with yellow centers that give off a faint roselike perfume. One sultry afternoon I watched the bumblebees making their rounds of my potato blossoms, thoughtlessly chalking themselves with yellow pollen grains before lumbering off to appointments with other blossoms, other species.

Uncertainty is the theme that unifies most of the questions now being raised about agricultural biotechnology by environmentalists and scientists. By planting millions of acres of genetically altered plants, we're introducing something novel into the environment and the food chain, the consequences of which are not completely understood. Several of these uncertainties have to do with the fate of the grains of pollen these bumblebees are carting off from my potatoes.

For one thing, that pollen, like every other part of the plant, contains Bt toxin. The toxin, which is produced by a bacterium that occurs naturally in the soil, is generally thought to be safe for humans, yet the Bt in genetically modified crops is behaving a little differently from the ordinary Bt that farmers have been spraying on their crops for years. Instead of quickly breaking down in nature, as it usually does, genetically modified Bt toxin seems to be building up in the soil. This may be insignificant; we don't know. (We don't really know what Bt is doing in soil in the first place.) We also don't know what effect all this new Bt in the environment may have on the insects we *don't* want to kill, though there are reasons to be concerned. In laboratory experiments scientists have found that the pollen from Bt corn is lethal to monarch butterflies. Monarchs don't eat corn pollen, but they do eat, exclusively, the leaves of milkweed (*Asclepias syriaca*), a weed that is common in American cornfields. When monarch caterpillars eat milkweed leaves dusted with Bt corn pollen, they sicken and die. Will this happen in the field? And how serious will the problem be if it does? We don't know.

What is remarkable is that someone thought to ask the question in the first place. As we learned during the glory days of the chemical paradigm, the ecological effects of changes to the environment often show up where we least expect to find them. DDT in its time was thoroughly tested and found to be safe and effective—until it was discovered that this unusually long-lived chemical travels through the food chain and happens to thin out the shells of birds' eggs. The question that led scientists to this discovery wasn't even a question about DDT, it was a question about birds: Why is the world's population of raptors suddenly collapsing? DDT was the answer. Hoping not to encounter that sort of surprise again, scientists are busy trying to imagine the sorts of questions to which Bt or Roundup Ready crops might someday prove to be the unexpected answer.

One of those questions has to do with "gene flow": What might happen to the Bt genes in the pollen my bumblebees are moving from blossom to blossom around my garden? Through cross-pollination those genes can wind up in other plants, possibly conferring a new evolutionary advantage on that species. Most domesticated plants do poorly in the wild; the traits we breed them for—fruit that ripens all at once, say—often render them less fit for life in the wild. But biotech plants have been given traits, such as insect or pesticide resistance, that render them *more* fit in nature.

Gene flow ordinarily occurs only between closely related species, and since the potato evolved in South America, the chances are slim that my Bt genes will escape into the wilds of Connecticut to spawn some kind of superweed. That's Monsanto's contention, and there's no reason to doubt it. But it is interesting to note that while genetic engineering depends for its power on the ability to break down the genetic walls between species and even phyla in order to freely move genes among them, the environmental safety of the technology depends on precisely the opposite phenomenon: on the integrity of species in nature and their tendency to reject alien genetic material.

Yet what will happen if Peruvian farmers plant Bt potatoes? Or if I plant a biotech crop that does have local relatives? Scientists have already proved that the Roundup Ready gene can migrate in a single generation from a field of rapeseed oil plants to a related weed in the mustard family, which then exhibits tolerance to the herbicide; the same has happened with genetically modified beets.

This came as no great surprise; what did is the discovery, in one experiment, that transgenes migrate more readily than ordinary ones; no one knows why, but these well-traveled genes may prove to be especially jumpy.

Jumping genes and superweeds point to a new kind of environmental problem: "biological pollution," which some environmentalists believe will be the unhappy legacy of agriculture's shift from a chemical to a biological paradigm. (We're already familiar with one form of biological pollution: invasive exotic species such as kudzu, zebra mussels, and Dutch elm disease.) Harmful as chemical pollution can be, it eventually disperses and fades, but biological pollution is self-replicating. Think of it as the difference between an oil spill and a disease. Once a transgene introduces a new weed or a resistant pest into the environment, it can't very well be cleaned up: it will already have become part of nature.

In the case of the NewLeaf potato, the most likely form of biological pollution is the evolution of insects resistant to Bt, a development that would ruin one of the safest insecticides we have and do great harm to the organic farmers who depend on it.* The phenomenon of insect resistance offers an object lesson in the difficulties of controlling nature, as well as the problem with using a linear machine metaphor to deal with a process as complex and nonlinear as evolution. For this is a case where the more thorough our control of nature is, the sooner natural selection will overthrow it.

According to the theory, which is based on classical Darwinism,

*What the emergence of Bt resistance might mean for the environment is harder to say. We have lots of experience with pests developing resistance to man-made pesticides, but what will happen if one of nature's own "pest controls" loses its effectiveness?

the new Bt crops add so much Bt toxin to the environment on such a continuous basis that the target pests will evolve resistance to it; the only real question is how long this will take to happen. Before now resistance hasn't been a worry, because conventional Bt sprays break down quickly in sunlight and farmers spray only when confronted with a serious infestation. Resistance is essentially a form of coevolution that occurs when a given population is threatened with extinction. That pressure quickly selects for whatever chance mutation will allow the species to change and survive. Through natural selection, then, one species' attempt at total control can engender its own nemesis.

I was surprised to learn that the specter of Bt resistance has forced Monsanto to temporarily lay aside its mechanistic habits of thought and approach the problem more like, well, a Darwinian. Working with government regulators, the company has developed a "Resistance Management Plan" to postpone Bt resistance. Farmers who plant Bt crops must leave a certain portion of their land planted in non-Bt crops in order to create "refuges" for the targeted bugs. The goal is to prevent the first Bt-resistant Colorado potato beetle from mating with a second resistant bug and thereby launching a new race of superbugs. The theory is that when that first Bt-resistant insect does show up, it can be induced to mate with a susceptible bug living on the refuge side of the tracks, thereby diluting the new gene for resistance. The plan implicitly acknowledges that if this new control of nature is to last, a certain amount of no-control, or wildness, will have to be deliberately cultivated. The thinking may be sound, but an awful lot has to go right for Mr. Wrong to meet Miss Right. No one can be sure how big the refuges have to be, where they should be located, and whether farmers will cooperate (creating safe havens for your most destructive pests is counterintuitive, after all)—not to mention the bugs.

Monsanto executives voice confidence that the plan will work, though their definition of success will come as small comfort to organic farmers: the company's scientists say that, if all goes well, resistance can be postponed for thirty years. After that? Dave Hjelle, the company's director of regulatory affairs, told me over lunch in St. Louis that Bt resistance shouldn't overly concern us since "there are a thousand other Bts out there"—that is, other proteins with insecticidal properties. "We can handle this problem with new products. The critics don't know what we have in the pipeline." This is, of course, how chemical companies have always handled the problem of pest resistance: by simply introducing a new and improved pesticide every few years. With any luck, the effectiveness of the last one will expire around the same time its patent does.

Behind the bland corporate assurances, though, stands a fairly startling admission. Monsanto is acknowledging that, in the case of Bt, it plans on simply using up not just another patented synthetic chemical but a natural resource, one that, if it belongs to anyone, belongs to everyone. The true cost of this technology is being charged to the future—no new paradigm there. Today's gain in control over nature will be paid for by tomorrow's new disorder, which in turn will become simply a fresh problem for science to solve. *We can cross that bridge when we come to it.* Of course, it was precisely this attitude toward the future that encouraged us to build nuclear power plants before anybody had figured out what to do with the waste—a bridge we now badly need to cross but find we still don't have any idea how to.

Dave Hjelle is a disarmingly candid man, and before we finished our lunch he uttered two words that I never thought I'd hear from the lips of a corporate executive, except perhaps in a bad movie. I'd assumed these two words had been scrupulously expunged from the corporate vocabulary many years ago, during a previous paradigm long since discredited, but Dave Hjelle proved me wrong:

"Trust us."

∷

July 7. My Colorado potato beetle vigil came to an end the first week of July, shortly before I planned to fly to Idaho to visit potato growers. I found a small platoon of larvae—soft brown dabs wearing what looked like miniature backpacks—munching with impunity on the leaves of my ordinary potato plants. I couldn't find a single one of the bugs on my NewLeafs, however, either alive or dead. Glenda Debrecht, the Monsanto horticulturist, had prepared me for this: predator insects were probably feasting on the bugs the NewLeafs had killed. I kept looking for them, though, and eventually I spied a single mature beetle sitting on a NewLeaf leaf; when I reached to pick it up, the beetle fell drunkenly to the ground. It had been sickened by the plant and would shortly be dead. My NewLeafs were working.

I have to admit to a certain thrill, a triumphal feeling that any gardener who has battled pests will understand. The typical gardener is not in the least bit romantic about the wildlife that assaults his plants, not the bugs or the woodchucks or the deer, and in his heart of hearts he believes that all is fair in war—even if organic principles (sort of like the Geneva Convention) do sometimes prevent him from heeding his heart's desire. But make no mistake, this is a desire whose fantasies feature rifles, explosives, and chemicals of unspeakable toxicity. So to watch a potato plant single-handedly vanquish a potato beetle is, at least from this perspective, a thing of beauty—an ingenious new twist on the Agricultural Sublime.

∷

Idaho, July 8. The Agricultural Sublime was very much on my mind during my flight west, especially as we crossed into Idaho. From thirty thousand feet, the perfect green circles formed by the irrigation pivots of the dryland farmer are breathtaking; in places the Idaho landscape becomes an endless grid of verdant coins pressed into the scrubby brown desert: squared circles as far as the eye can see. It's an image not only of human order, like the rows of corn back home, but also, in a landscape as inhospitable as the American West, of human habitation hard-won. I would soon discover, however, that this austere beauty is harder to see on the ground.

No one can make a better case for a biotech crop than a potato farmer, which is why Monsanto was eager for me to come out to Idaho to meet a few of their customers. From where a typical American potato grower stands, the NewLeaf looks very much like a godsend. That's because the typical potato grower stands in the middle of a bright green circle of plants that have been doused with so much pesticide that their leaves wear a dull white chemical bloom and the soil they're rooted in is a lifeless gray powder. Farmers call this a "clean field," since, ideally, it has been cleansed of all weeds and insects and disease—of all life, that is, with the sole exception of the potato plant. A clean field represents a triumph of human control, but it is a triumph that even many farmers have come to doubt. To such a farmer a new kind of potato that promises to eliminate the need for even a single spraying of chemicals is, very simply, an economic and environmental and perhaps even psychological boon.

Danny Forsyth laid out the chemistry and economics of modern potato growing for me one sweltering morning at the sleepy but well-air-conditioned coffee shop in Jerome, Idaho, a one-street, one-coffee-shop town about a hundred miles east of Boise on the interstate. Forsyth is a slight, blue-eyed man in his early sixties with a small, unexpected gray ponytail, a somewhat nervous manner, and a passing resemblance to Don Knotts. He farms three thousand acres of potatoes, corn, and wheat here in the Magic Valley, much of it on land inherited from his father. When he talks about agricultural chemicals, he sounds like a man desperate to kick a bad habit.

"None of us would use them if we had any choice," he said; he believes Monsanto is offering him that choice.

I asked Forsyth to walk me through a season's regimen, the state of the art in the control of a potato field. Typically it begins early in the spring with a soil fumigant; to control nematodes and certain diseases in the soil, potato farmers douse their fields before planting with a chemical toxic enough to kill every trace of microbial life in the soil. Next Forsyth puts down an herbicide—Lexan, Sencor, or Eptam—to "clean" his field of all weeds. Then, at planting, a systemic insecticide—such as Thimet—is applied to the soil. This will be absorbed by the young seedlings and kill any insect that eats their leaves for several weeks. When the potato seedlings are six inches tall, a second herbicide is sprayed on the field to control weeds.

Dryland farmers like Forsyth farm in the vast circles I'd seen from the sky; each circle, defined by the radius of the irrigation pivot, typically covers an area of 135 acres. Pesticides and fertilizer are simply added to the irrigation system, which on Forsyth's farm draws water from (and returns it to) the nearby Snake River. Along with their ration of water, Forsyth's potatoes receive ten weekly sprayings of chemical fertilizer. Just before the rows close—when the leaves of one row of plants meet those of the next—he begins spraying Bravo, a fungicide, to control late blight, the same fungus that caused the Irish potato famine and is once again today the potato grower's most worrisome threat. A single spore can infect a field overnight, Forsyth said, turning the tubers into a rotting mush.

Beginning this month, Forsyth will hire a crop duster to spray for aphids at fourteen-day intervals. The aphids are harmless in themselves, but they transmit the leaf roll virus, which causes "net necrosis" in Russet Burbanks, a brown spotting of the potato's flesh that will cause a processor to reject a whole crop. Despite all his efforts to control it, this happened to Forsyth just last year. Net necrosis is a purely cosmetic defect, yet because McDonald's believes—with good reason—that we don't like to see brown spots in our french fries, farmers like Danny Forsyth must spray their fields with some of the most toxic chemicals now in use, including an organophosphate called Monitor.

"Monitor is a deadly chemical," Forsyth told me; it is known to damage the human nervous system. "I won't go into a field for four or five days after it's been sprayed—not even to fix a broken pivot." That is, Forsyth would sooner lose a whole circle to drought than expose himself or an employee to this poison.

Leaving aside the health and environmental costs, the economic cost of all this control is daunting. A potato farmer in Idaho spends roughly $1,950 an acre (mainly on chemicals, electricity, and water) to grow a crop that, in a *good* year, will earn him maybe $2,000. That's how much a french-fry processor will pay for the twenty tons of potatoes a single Idaho acre can yield. It's not hard to see why a farmer like Forsyth, struggling against such tight margins and heartsick over chemicals, would leap at a NewLeaf.

"The NewLeaf means I can skip a couple of sprayings," Forsyth said. "I save money, and I sleep better. It also happens to be a nice-looking spud."

Before driving out to have a look at his fields, Forsyth and I got onto the subject of organic agriculture, about which he had the usual things to say ("That's all fine on a small scale, but they don't have to feed the world") and a few things I never expected to hear from a conventional farmer. "I like to eat organic food, and in fact I grow a lot of it at the house. The vegetables we buy at the market we just wash and wash and wash. I'm not sure I should be saying this, but I always plant a small area of potatoes without any chemicals. By the end of the season, my field potatoes are fine to eat, but any potatoes I pulled today are probably still full of systemics. I don't eat them."

Danny Forsyth's words came back to me a few hours later, during lunch at the home of another Magic Valley farmer. Steve Young is a progressive, prosperous potato grower—"a player" in the admiring words of my Monsanto escort. A big, bluff man in his forties, Young farms ten thousand acres; even after passing the entrance to his farm, you have to drive for miles before arriving at his house. He showed me the computers that automatically regulate his eighty-five circles of spuds; each circle on the screen stands for and controls one circle in the field. Without so much as stepping outside, Young can water his fields or spray them with pesticide. Young seemed very much the master of his fate as well as his fields, the picture of a thoroughly modern farmer. He's built his own potato storage facility—a controlled-atmosphere shed big as a football field, housing a mountain of Russet Burbanks thirty feet tall—and he owns a share in a local chemical distributorship. Compared to Danny Forsyth, a man who clearly feels himself very much at the mercy of the chemicals, the aphids, and the potato processors, Young gives the impression, at least, of a man in complete control.

Mrs. Young had prepared a lavish feast for us, and after Dave, their eighteen-year-old, said grace, adding a special prayer for me (the Youngs are devout Mormons), she passed around a big bowl of potato salad. As I helped myself, my Monsanto escort asked her what was in the salad, flashing me a smile that suggested she might already know.

"It's a combination of NewLeafs and some of our regular Russets," Mrs. Young said, positively beaming. "Dug this very morning."

:|:

As I slowly chewed my potato salad, I considered which ingredient was more likely to be hazardous to my health, the NewLeafs or the Russets à la Thimet? The answer, I decided, is almost certainly potato number two. There might be unknowns about the NewLeafs, but the Russets I *knew* to be full of poison—and the answer says something important about genetically engineered plants I wasn't ready to hear, at least not before coming out to Idaho. After I talked to farmers like Danny Forsyth and Steve Young while walking fields made sterile by a drenching, season-long rain of chemicals, Monsanto's NewLeafs began to look like a blessing. Set against current practices, genetically modified potatoes represent a more sustainable way of growing food. The problem is, that isn't saying much.

After my lunch with the Youngs, I shook off my escort long enough to pay a visit to a nearby organic potato grower. I knew enough not to take someone from Monsanto to visit an organic farm. "If there's a source of evil in agriculture," an organic farmer from Maine had told me, "its name is Monsanto."

Mike Heath is a rugged, lined, laconic man in his mid-fifties. Like most of the organic farmers I've ever met, he looks as though he spends a lot more time outdoors than a conventional farmer, and he probably does: chemicals are, among other things, labor-saving devices. While we drove around his five hundred acres in a battered old pickup truck, I asked him what he thought about genetic engineering. He voiced many reservations—it was synthetic, there were too many unknowns—but his main objection to planting a biotech potato was simply that "it's not what my customers want."

I asked Heath about the NewLeaf potato. He had no doubt that resistance would come—"Face it," he said, "the bugs are always going to be smarter than we are"—and he felt it was unjust that Monsanto was profiting from the ruin of a "public good" such as Bt.

None of this particularly surprised me; what did was the fact that Heath himself had resorted to spraying Bt on his potatoes only once or twice in the last ten years. I had assumed that organic farmers used Bt and the other approved pesticides in much the same way conventional farmers use theirs, but as Mike Heath showed me around his farm, I began to understand that organic farming was a lot more complicated than simply substituting good inputs for bad. A whole different metaphor seemed to be involved.

Instead of buying many inputs at all, Heath relies on a long and complex crop rotation to avoid a buildup of crop-specific pests. He's found, for instance, that planting wheat in a field prior to potatoes "confuses" the potato beetles when they emerge from their larval stage. He also plants strips of flowering plants on the margins of his potato fields—peas or alfalfa, usually—to attract the beneficial insects that dine on beetle larvae and aphids. If there aren't enough beneficial insects around to do the job, he'll introduce ladybugs. Heath also grows a dozen different varieties of potatoes, on the theory that biodiversity in a field, as in the wild, is the best defense against nature's inevitable surprises. A bad year with one variety will likely be offset by a good year with the others. He doesn't, in other words, ever bet the farm on a single crop.

By way of driving home a point, Heath dug some of his Yukon Golds for me to take home. "I can eat any potato in this field right now. Most farmers can't eat their spuds out of the field." I decided not to mention my lunch.

For fertilizers, Heath relies on "green manures" (growing cover crops and plowing them under), cow manure from a local dairy, and the occasional spraying of liquefied seaweed. The result was a soil that looked completely different from the other Magic Valley soils I'd fingered that day: instead of the uniform grayish powder I'd assumed was normal for the area, Heath's soil was dark brown and crumbly. The difference, I understood, was that this soil was alive. Much more than an inert mechanism for conducting water and chemicals to the crop's roots, it actually contributed nutrients of its own making to the plants. The biology, chemistry, and physics of this process, which goes by the name "fertility," is not at all well understood—soil truly is a wilderness—yet this ignorance doesn't prevent organic farmers and gardeners from nurturing it.

Heath's crops looked different, too: more compact plants (chemical fertilizers tend to make plants leafier); the occasional weed, and loads of insects flitting around. Here were the very opposite of "clean" fields, and, frankly, their weedy hedgerows and overall patchiness made them much less pretty to look at. To the eye, at least, the order of these fields seemed much softer and less complete, with a great deal of disorder percolating at the margins. Of course, what the eye failed to see was a more complex, less human order—the order, that is, of an ecosystem, one that is not so much imposed by the farmer as it is nourished and tweaked by him. It is the very complexity of such fields—the sheer diversity of species in both space and time—that makes them productive year after year without many inputs. The system provides for most of its needs.

On the drive back to Boise, I thought about why Mike Heath's farm remains the exception, both in Idaho and elsewhere. Here was a genuinely new paradigm—a biological paradigm—and it seemed to work: Heath spends a fraction as much on inputs as Danny Forsyth or Steve Young, yet he was digging between three and four hundred bags per acre—just as many as Forsyth and only slightly fewer than Young.* But while organic agriculture is gaining ground, few of the mainstream farmers I met considered it a "realistic" alternative to the way we presently grow our food.

They may be right. In a dozen different ways, a farm like Mike Heath's simply can't be reconciled to the logic of a corporate food chain. For one thing, Heath's type of agriculture doesn't leave much room for the Monsantos of this world: organic farmers buy remarkably little—some seed, a few tons of manure, maybe a few gallons of ladybugs. The organic farmer's focus is on a process, rather than on products. Nor is that process readily systematized, reduced to, say, a prescribed regimen of sprayings like the one Danny Forsyth laid out for me—regimens that are typically designed by companies selling chemicals. Most of the intelligence and local knowledge needed to run Mike Heath's farm resides in the head of Mike Heath. Growing potatoes conventionally re-

---

*Before Heath's operation can be compared to a conventional farm, you have to factor in the additional labor (many smaller crops mean more work; organic fields must also be cultivated for weeds) and time—the typical organic rotation calls for potatoes every fifth year, rather than every third as on a conventional farm. Even so, Heath gets almost twice the price for his spuds: $9.00 a bag from an organic processor that ships frozen french fries to Japan.

quires intelligence, too, but a larger portion of it resides in laboratories in distant places such as St. Louis, where it is employed developing inputs like Roundup or the NewLeaf.

This sort of centralization of agriculture is not likely to be reversed any time soon, if only because there's so much money in it and, in the short run at least, it's so much easier for the farmer to buy prepackaged solutions from big companies. "Whose Head Is the Farmer Using?" asks the title of a Wendell Berry essay; "Whose Head Is Using the Farmer?" At a certain point, a point already long past, the farmer's attempt at the perfect control of nature evolved into the control of the farmer by the corporations that promoted that dream in the first place. It is only because that dream is so elusive that the control of farmers by its merchants became so inescapable.

:‖:

Organic farmers like Mike Heath have turned their backs on what is unquestionably the greatest strength—and still greater weakness—of industrial agriculture: monoculture and the economies of scale it makes possible. Monoculture is the single most powerful simplification of modern agriculture, the key move in reconfiguring nature as a machine, yet nothing else in agriculture is so poorly fitted to the way nature seems to work. Very simply, a vast field of identical plants will always be exquisitely vulnerable to insects, weeds, and disease—to all the vicissitudes of nature. Monoculture is at the root of virtually every problem that bedevils the modern farmer, and from which virtually every agricultural product is designed to deliver him.

To put the matter baldly, a farmer like Mike Heath is working hard to adjust his fields to the logic of nature, while Danny Forsyth is working even harder to adjust his fields to the logic of monoculture and, standing behind that, the logic of an industrial food chain. One small case in point: when I asked Mike Heath what he did about net necrosis, the bane of Danny Forsyth's potato crop, I was disarmed by the simplicity of his answer. "That's only really a problem with Russet Burbanks," he explained. "So I plant other kinds." Forsyth can't do that. He's part of a food chain—at the far end of which stands a perfect McDonald's french fry—that demands he grow Russet Burbanks and nothing else.

This, of course, is where biotechnology comes in, to the rescue of Forsyth's Russet Burbanks and, Monsanto is betting, to the whole industrial food chain of which they form a part. Monoculture is in crisis. The pesticides that make it possible are rapidly being lost, either to resistance or to worries about their dangers. As the fertility of the soil has declined under the onslaught of chemicals, so too in many places have crop yields. "We need a new silver bullet," an entomologist with the Oregon Extension Service told me, "and biotech is it." Yet a new silver bullet is not the same thing as a new paradigm. Rather, it's something that will allow the old paradigm to survive. That paradigm will always construe the problem in Danny Forsyth's field as a Colorado beetle problem, rather than what it is: a problem of potato monoculture.

⁝⁝

What Mike Heath's answer to my question about net necrosis— "That's only really a problem with Russet Burbanks"—suggests is that the problem of monoculture may itself be as much a problem of *culture* as it is of agriculture. Which is to say, it's a problem in which all of us are implicated, not just farmers and companies like Monsanto. I was starting to appreciate that the conventional journalistic narrative that usually organizes a story like this—evil technology foisted by greedy corporation—leaves out an important element, which is us and *our* desire for control and uniformity. So much of what I'd seen in Idaho—from the clean fields to the computer-controlled crop circles—goes back to that perfect McDonald's french fry at the eating end of the food chain.

On my way back to Boise I did a drive-through at a McDonald's and ordered a bag of the fries in question. There's no way of knowing for sure, but these fries may well have been my second meal of NewLeafs in a day; at the time, McDonald's used NewLeafs in its french fries. A Monsanto executive had told me that without McDonald's early support the NewLeaf might never have gotten off the ground, since McDonald's is one of the largest buyers of potatoes in the world.*

You know, their fries really are gorgeous: slender golden rectangles long enough to overshoot their trim red containers like a bouquet. A farmer had told me that only the Russet Burbank will give you a fry quite that long and perfect. To look at them is to appreciate that these aren't *just* french fries: they're Platonic ideals of french fries, the image and the food rolled into one, and available anywhere in the world for around a dollar a bag. You can't beat it.

I wanted and fully expected to find precisely the same Platonic french fry here in Nowhere, Idaho, that I'd had countless times at home and could expect to find anytime I wanted to in Tokyo, Paris, Beijing, Moscow, even Azerbaijan or the Isle of Man. What is that, if not a control thing?—and not just on the part of

---

*Recently, McDonald's and several other large food companies, responding to growing public unease about genetically modified food, have stopped using genetically modified crops in their products. The loss of McDonald's support appears to have doomed the NewLeaf potato: Monsanto announced in early 2001 that the product would be discontinued the following year. Acreage planted to other GMO crops continues to increase, though at a slower rate.

McDonald's. But whatever is behind it, this expectation can't be fulfilled unless McDonald's has seen to it that millions of acres of Russet Burbanks are planted all over the world. The global desire can't be gratified without the global monoculture, and that global monoculture now depends on technologies like genetic engineering. It just may be that we can't have the one without the other.

This alignment of global desire and technology has been a great boon for the Russet Burbank, at least in terms of sheer numbers. Has there ever been a more successful potato in the history of the world? Yet its success is a precarious thing, for this particular set of potato genes (or rather now, potato genes plus one Bt gene and one antibiotic-resistance gene, courtesy of Monsanto) has also never been more vulnerable to the vicissitudes of nature or the fecklessness of a single species: us. Whether in evolutionary terms a monoculture really represents long-term success for a species is an open question. The Lumper, Ireland's favorite potato before the famine, was once nearly as dominant as the Russet Burbank; today, its genes are as hard to find as the dodo's.

Part of the pleasure those fries gave me was how perfectly they conformed to my image and expectation of them—to the Idea of Fries in my head, that is, an idea that McDonald's has successfully planted in the heads of a few billion other people around the world. Here, then, is a whole other meaning of the word *monoculture*. Like the agricultural practice that goes by that name, this one too—the monoculture of global taste—is about uniformity and control. Indeed, the monocultures of the field and the monocultures of our global economy nourish each other in crucial ways. The two are complexly intertwined expressions of the same Apollonian desire, our impulse, I mean, to elevate the universal over the particular or local, the abstract over the concrete, the ideal over the real, the made over the natural. The spirit of Apollo celebrates "the One," Plutarch wrote, "denying the many and ab-juring multiplicity." Against Dionysus's "variability" and "wan-tonness" he poses the power of "uniformity [and] orderliness." Apollo is the god, then, of monoculture, whether of plants or of people. And though Apollo has surely had many more exalted manifestations than this one, he is here, too, in every bag of McDonald's french fries.

:||:

Ireland, 1846. "On the 27th of last month [July] I passed from Cork to Dublin, and this doomed plant bloomed in all luxuriance of an abundant harvest." So begins a letter written in the summer of 1846 by a Catholic priest named Father Mathew. "Returning on the 3rd [of August] I beheld with sorrow one wide waste of putre-fying vegetation. In many places the wretched people were seated on the fences of their decaying gardens, wringing their hands, and wailing bitterly the destruction that had left them foodless."

The arrival of the blight was announced by the stench of rot-ting potatoes, a stench that became general in Ireland late in the summer of 1845, then again in '46 and '48. Its spores carried on the wind, the fungus would appear in a field literally overnight: a black spotting of the leaves followed by a gangrenous stain spread-ing down the plant's stem; then the blackened tubers would turn to evil-smelling slime. It took but a few days for the fungus to scorch a green field black; even potatoes in storage succumbed.

The potato blight visited all of Europe, but only in Ireland did it produce a catastrophe. Elsewhere, people could turn to other staple foods when a crop failed, but Ireland's poor, subsisting on potatoes and exiled from the cash economy, had no alternative. As is often the case in times of starvation, the problem was not quite so simple as a shortage of food. At the height of the famine, Ireland's docks were heaped with sacks of corn destined for export to England. But the corn was a commodity, determined to follow the money; since the potato eaters had no money to pay for corn, it sailed for a country that did.

The potato famine was the worst catastrophe to befall Europe since the Black Death of 1348. Ireland's population was literally decimated: one in every eight Irishmen—a million people—died of starvation in three years; thousands of others went blind or insane for lack of the vitamins potatoes had supplied. Because the poor laws made anyone who owned more than a quarter acre of land ineligible for aid, millions of Irish were forced to give up their farms in order to eat; uprooted and desperate, the ones with the energy and wherewithal emigrated to America. Within a decade, Ireland's population was halved and the composition of America's population permanently altered.

Contemporary accounts of the potato famine read like visions of Hell: streets piled with corpses no one had the strength to bury, armies of near-naked beggars who'd pawned their clothes for food, abandoned houses, deserted villages. Disease followed on famine: typhus, cholera, and purpura raced unchecked through the weakened population. People ate weeds, ate pets, ate human flesh. "The roads are beset with tattered skeletons," one witness wrote. "God help the people."

The causes of Ireland's calamity were complex and manifold, involving such things as the distribution of land, brutal economic exploitation by the English, and a relief effort by turns heartless and hapless, as well as the usual accidents of climate, geography, and cultural habit. Yet this whole edifice of contingency rested at bottom upon a plant—or, more precisely, upon the relationship between a plant and a people. For it was not the potato so much as potato monoculture that sowed the seeds of Ireland's disaster.

Indeed, Ireland's was surely the biggest experiment in monoculture ever attempted and surely the most convincing proof of its folly. Not only did the agriculture and diet of the Irish come to depend utterly on the potato, but they depended almost completely on one kind of potato: the Lumper. Potatoes, like apples, are clones, which means that every Lumper was genetically identical to every other Lumper, all of them descended from a single plant that just happened to have no resistance to *Phytophthora infestans.* The Incas too built a civilization atop the potato, but they cultivated such a polyculture of potatoes that no one fungus could ever have toppled it. In fact, it was to South America that, in the aftermath of the famine, breeders went to look for potatoes that could resist the blight. And there, in a potato called the Garnet Chile, they found it.

Monoculture is where the logic of nature collides with the logic of economics; which logic will ultimately prevail can never be in doubt. In Ireland under British rule the logic of economics dictated a monoculture of potatoes; in 1845, the logic of nature exercised its veto, and a million people—many of whom probably owed their existence to the potato in the first place—perished.

"As for their command over Nature," wrote Benjamin Disraeli in his 1847 novel *Tancred,* "let us see how it will operate in a second deluge. Command over nature! Why the humblest root that serves for the food of man has mysteriously withered throughout Europe, and they are already pale at the possible consequences."

∷

In March 1998, patent number 5,723,765, describing a novel method for the "control of plant gene expression," was granted jointly to the U.S. Department of Agriculture and a cottonseed company called Delta & Pine Land. The bland language of the patent obscures a radical new genetic technology: introduced into any plant, the gene in question causes the seeds that plant makes to become sterile—to no longer do what seeds have always done. With the "Terminator," as the new technique quickly became known, genetic engineers have discovered how to stop on command the most elemental of nature's processes, the plant-seed-plant-seed cycle by which plants reproduce and evolve. The ancient logic of the seed—to freely make more of itself ad infinitum, to serve as both food and the means of making more food in the future—has yielded to the modern logic of capitalism. Now viable seeds will come not from plants but from corporations.

The dream of controlling the seed, and through the seed the farmer, is older than genetic engineering. It goes back at least to the development, in a handful of crops, of modern hybrids, high-yielding varieties that don't "come true" from replanted seed, thereby forcing farmers to buy new seeds every spring. Yet compared to the rest of the economy, farming has largely resisted the trend toward centralization and corporate control. Even today, when only a handful of big companies are left standing in most American industries, there are still some two million farmers. What has stood in the way of concentration is nature: her complexity, diversity, and sheer intractability in the face of our most heroic efforts at control. Perhaps most intractable of all has been agriculture's means of production, which of course is nature's own: the seed.

It's only in the last few decades, with the introduction of modern hybrids, that farmers began to buy their seeds from big companies. Even today a great many farmers save some seed every fall to replant in the spring. "Brown bagging," as this practice is sometimes called, allows farmers to select strains particularly well adapted to local conditions.* Since these seeds are typically traded

---

*Worldwide, it's estimated that some 1.4 billion people depend on saved seed.

among farmers, the practice steadily advances the state of the genetic art. Indeed, over the centuries it has given us most of our major crop plants.

Infinitely reproducible, seeds by their very nature don't lend themselves to commodification, which is why the genetics of most of our major crop plants have traditionally been regarded as a common heritage rather than as "intellectual property." In the case of the potato, the genetics of the important varieties—the Russet Burbanks and Atlantic Superiors, the Kennebecs and Red Norlings—have always been in the public domain. Before Monsanto got involved, there had never been a national corporation in the potato seed business. There simply wasn't enough money in it.

Genetic engineering changes this. By adding a new gene or two to a Russet or Superior, Monsanto can now patent the improved variety. Legally, it's been possible to patent a plant for several years now, but biologically, these patents have been almost impossible to enforce. Genetic engineering has gone a long way toward solving this problem, since it allows Monsanto to test the potato plants growing on a farm to prove they're the company's intellectual property. The contracts farmers must sign to buy Monsanto seeds grant the company the right to perform such tests at will, even in future years. To catch farmers violating its patent rights, Monsanto has reportedly paid informants and hired Pinkertons to track down gene thieves; it has already sued hundreds of farmers for patent infringement. With a technology such as the Terminator, the company will no longer have to go to all that trouble.*

---

*I say "such as the Terminator" because, after an international barrage of criticism, Monsanto has forsworn the technology. However, it has not forsworn a group of related technologies that achieve the same end: Genetic Use Restriction Technologies (GURT), which make it possible to turn genetic traits on and off by

With the Terminator, seed companies can enforce their patents biologically and indefinitely. Once these genes are widely introduced, control over the genetics of our crop plants and the trajectory of their evolution will complete its move from the farmer's field to the seed company—to which the world's farmers will have no choice but to return year after year. The Terminator allows companies like Monsanto to enclose one of the last great commons in nature: the genetics of the crop plants that civilization has developed over the past ten thousand years.

At lunch I had asked Steve Young what he thought about all this, especially about the contract Monsanto forces him to sign and the prospect of sterile seeds. I wondered how the American farmer, the putative heir to a long tradition of agrarian independence, was adjusting to the idea of field men snooping around his farm and patented seeds he couldn't replant.

Young told me he'd made his peace with corporate agriculture, and with biotechnology in particular. "It's here to stay. It's necessary if we're going to feed the world, and it's going to take us forward."

I asked him if he saw any downside to biotechnology. Someone from Monsanto was with us at the table; Young's reply was a long time in coming, and the moment grew uncomfortable. What he finally said silenced the table, and made me think again about the image of mastery he'd projected—about the computer-controlled fields, the chemical distributorship, the miles of patented high-tech spuds framed in his living room's picture window, reaching clear to the horizon.

---

applying certain proprietary chemicals to genetically modified plants in the field. So even if the plant in question still produces viable seed, those seeds will produce worthless plants—plants with their disease or herbicide resistance turned off—unless the farmer buys the chemical activator.

"Oh, there is a cost all right," Young said darkly. "It gives corporate America one more noose around my neck."

⁝⁝

August. A few weeks after I got home from Idaho, I dug my NewLeafs, harvesting a gorgeous-looking pile of spuds, several real lunkers among them. The plants had performed brilliantly, though so had all my other potatoes: the beetle problem never got out of hand, perhaps because the diversity of species in my garden had attracted enough beneficial insects to check the bugs. Who knows? My scapegoat tomatillos may also have helped. The fact is, a true test of my NewLeafs would have meant planting a monoculture.

By the time I harvested my crop, the question of eating my NewLeafs was moot. Whatever I thought about the safety of these potatoes really didn't matter. Not just because I'd already tasted Mrs. Young's NewLeaf potato salad but because Monsanto and my government had long ago taken the decision as to whether or not to eat a biotech spud out of my hands. Chances are I've eaten plenty of NewLeafs already, at McDonald's or in bags of Frito-Lay chips, though without a label, there's no way of knowing for sure.

So if I've eaten probable NewLeafs already, why was it I kept putting off eating these definite NewLeafs? Maybe just because it was August and there were so many more interesting fresh potatoes around—fingerlings with dense, luscious flesh, Yukon Golds (Mike Heath's as well as my own) that looked and tasted as though they'd been buttered in the skin—that the idea of cooking with the sort of bland commercial variety Monsanto puts its genes into seemed almost beside the point.

There was this, too: I'd called some of the government agencies in Washington that had signed off on the NewLeaf, and what they said didn't exactly fill me with confidence. The Food and Drug

Administration told me that, because it operates on the assumption that genetically modified plants are "substantially equivalent" to ordinary plants, the regulation of these foods has been voluntary since 1992. Only if Monsanto feels there is a safety concern is it required to consult with the agency about its NewLeafs. I'd always assumed the FDA had tested the new potato, maybe fed a bunch of them to rats, but it turned out this was not the case. In fact, the Food and Drug Administration doesn't even officially regard the NewLeaf as a food. *What?* It seems that since the potato contains Bt, it is, at least in the eyes of the federal government, not a food at all but a pesticide, putting it under the jurisdiction of the Environmental Protection Agency. Feeling a bit like Alice in a bureaucratic wonderland, I phoned the EPA to ask about my potatoes. As the EPA sees it, Bt has always been a safe pesticide, the potato has always been a safe food, so put the two together and you've got something that should be safe both to eat and to kill bugs with. Evidently the machine metaphor has won the day in Washington too: the NewLeaf is simply the sum of its parts—a safe gene added to a safe potato.

I also phoned Margaret Mellon at the Union of Concerned Scientists in Washington, D.C., to ask her advice about my spuds. Mellon is a molecular biologist and lawyer and a leading critic of biotech agriculture. She couldn't offer any hard scientific proof that my NewLeafs were unsafe to eat, but she pointed out that there was also no scientific proof for the notion of "substantial equivalence."*

"That research simply hasn't been done."

---

*In fact, internal documents that have come to light as part of a consumer suit against the FDA reveal that several of the agency's own scientists also reject the notion of "substantial equivalence."

Mellon talked about genetic instability, a phenomenon which strongly suggests that a biotech plant is *not* simply the sum of its old and new genes, and she talked about the fact that we know nothing about the effect of Bt in the human diet, a place it has never been before. I pressed: Was there any reason why I shouldn't eat these spuds?

"Let me ask *you* a question: Why would you want to?"

This was a good question. So for several weeks late that summer my NewLeafs remained in a shopping bag on the porch. Then I took the bag with me on vacation, thinking maybe I'd sample them there, but the bag came home untouched—except, that is, for one potato I'd taken out of it. A fishmonger had told me about a Martha Stewart tip for keeping grilled fish from sticking to a barbecue: rub the grill with a raw potato sliced down the middle. It works, by the way.

But I was still left with my bag of NewLeafs sitting there on the porch. And there they sat until Labor Day, when I got an invitation to a potluck supper at the town beach. Perfect! I signed up to make a potato salad. The day of the supper, I brought the bag of spuds into the kitchen and set a pot of water on the stove. But before the water even had a chance to boil, I was stricken by this obvious thought: Wouldn't I have to tell people at the picnic what they were eating? I had no reason to think the potatoes weren't perfectly safe, but if the idea of eating genetically modified food without knowing it gave me pause, I couldn't very well ask my neighbors to do so. (That would be rather more potluck than they were counting on.) So of course I'd have to tell them all about the NewLeafs—and then, no doubt, bring home a big bowl of untouched potato salad. For surely there'd be other potato salads at the potluck, and who, given the choice, was ever going to opt for the one with the biotech spuds? I suddenly understood with perfect clarity why Monsanto doesn't want to label its genetically modified food.

So I turned down the flame under the pot and went out to the garden to harvest a pile of ordinary spuds for my potato salad. The NewLeafs went back out into the limbo of my porch.

I hadn't been in the garden for a couple of weeks, and, as always is the case by the end of the summer, the place was an anarchy of rampant growth and ripe fruit, all of it threatening to burst the geometry of my beds and trellises and paths. The pole beans had climbed clear to the tops of the sunflowers, which stood draped in their bulging green and yellow pods. The pumpkins had trailed halfway across the now-unmowable lawn, and the squash leaves, big as pizzas, threw dark pools of shade in which the lettuces looked extremely happy—as, unfortunately, did the slugs, who were dining on my chard in the squashy shade. The vines of the last potatoes lay flopped over their hills, exhausted.

The garden had come to this, had reached this pitch of green uproar in the few short weeks since May, when I'd set out seedlings in a considered pattern I no longer could discern. The neat, freshly hoed rows had once implied that I was in charge here, the gardener in chief, but clearly this was no longer the case. My order had been overturned as the plants went blithely about their plant destinies. This they were doing with the avidity of all annuals, reaching for the sun, seizing ground from neighbors, fending off or exploiting one another whenever the opportunity arose, ripening the seeds that would bear their genes into the future, and generally making the most of the dwindling days till frost.

For a while every season, I do try to keep the whole thing under some semblance of control, pulling the weeds, clipping back the squash so that the chard might breathe, untangling the bean vines before they choke their frailer neighbors. But by the end of August I usually give it up, let the garden go its own way while I simply try to keep up with the abundance of the late-summer harvest. By this point what's going on in the garden is no longer my doing, even if it was I who got the whole thing rolling back in May. As much as I love the firm grasp and cerebral order of spring, there's a ripe, almost sensual pleasure in its August abandonment, too.

But I'd come here looking for something, and eventually I found it: a row of Kennebecs, their tops already sprawled dead on the ground. One of the many virtues of potatoes is that they can be left in the ground all winter, to be dug only as needed; historically, this has been a great blessing to peasants subject to marauding armies, since potatoes in the ground can't very well be ransacked.

I think there is no harvest more satisfying than the harvest of potatoes. I love the moment when the spade turns over the crust of black soil for the first time since spring and the chino-colored lumps tumble out onto the fresh dirt. After gathering up the first flush of easy ones, you should lay the spade aside (or else you're apt to bruise the remaining potatoes). Go for the rest by hand, forcing your fingers down into the richly manured soil, feeling around in the dark for those unmistakable forms, the identity of which the hand has no need of the eye to confirm. That's because potatoes are always cooler to the touch than stone, heavier too, and somehow always a happier fit in the hand.

Not that any given spud is ever such a paragon of form. No two of them ever alike, most potatoes are odd, misshapen, asymmetrical things, their shapes determined as much by the accidents of adjacent rocks and soil as by any genetic instruction followed to the letter. Maybe this is why we like to give our chthonic spuds such sunny and Apollonian forms, slicing them into translucent chips and geometrical fries. Yet compared to the undifferentiated night in which they grow, the bright potatoes feel in the hand like form incarnate.

Sooner or later your fingers close on that one moist-cold spud that the spade has accidentally sliced clean through, shining wetly white and giving off the most unearthly of earthly aromas. It's the smell of fresh soil in the spring, but fresh soil somehow distilled or improved upon, as if that wild, primordial scene had been refined and bottled: *eau de pomme de terre*. You can smell the cold inhuman earth in it, but there's the cozy kitchen too, for the smell of potatoes is, at least by now, to us, the smell of comfort itself, a smell as blankly welcoming as spud flesh, a whiteness that takes up memories and sentiments as easily as flavors. To smell a raw potato is to stand on the very threshold of the domestic and the wild.

Once I'd filled a basket with my spuds, I stood and considered the state of the garden, the daunting magnificence of its declension from May's straightforward rows and intentions. Whenever I hear or read the word *garden*, I always picture something so much less wild than this, probably because in common usage *garden* stands as the opposite of *wilderness*. The gardener knows better than to believe this, though. He knows that his garden fence and path and cherished geometries hold in their precarious embrace, if not a wilderness in any literal sense, then surely a great, teeming effulgence of wildness—of plants and animals and microbes leading their multifarious lives, proposing so many different and unexpected answers to the deep pulse of their genes and the wide press of their surroundings—of everything affecting everything else.

So where exactly does that leave us—the gardeners and descendants of Johnny Appleseed who would try to make something of this wildness? Standing amid this sweet wreck of a garden this August afternoon, lifting a basket heavy with potatoes, I thought about Chapman in his coffee sack, about the fanatical tulip fanciers and pot growers of Amsterdam, about the Monsanto scientists in their lab coats, and wondered what they had in common. All of them had ventured into the garden—into Darwin's Ever-Expanding Garden of Artificial Selection—for the purpose of marrying powerful human drives to the equally powerful drives of plants; all were practitioners of the botany of desire. In the nature of things, this made them—Chapman-like, potatolike—figures of the margins, moving between the realms of the wild and the cultivated, the anciently given and the newly made, the Dionysian and the Apollonian. All of them had taken part in the great, never-to-be-concluded conversation between those two presiding deities, adding their two cents to the dialogue of Dionysian energy and Apollonian order that has produced the beauty of a Queen of Night tulip, the sweetness of a Jonagold apple, the perception sponsored in a human brain by *Cannabis sativa* × *indica*.

Somewhere between those two poles, all gardeners—indeed, all of us—stake out their ground, some of them, like Appleseed, leaning to the side of Dionysian wildness (he'd love this garden now); others, like the scientists at Monsanto, pushing toward the Apollonian satisfactions of control. (The lab coats would probably have liked the garden better earlier in the season, before all hell broke loose.) Still others are harder to place on the continuum: I mean, where exactly do you put the marijuana grower tending his hydroponic closet of clones—that Apollonian edifice dedicated to the pursuit of Dionysian pleasure? It's a good thing one doesn't have to take sides.

With the exception of John Chapman, who had the imagination to identify with the bees, all these other botanists of desire went about their work from a straightforward and, it seems to me, blinkered humanist perspective. They took it for granted that domestication was something people did to plants, never the other way around. It probably never occurred to Dr. Adriaen Pauw, the Dutch burgher who owned eleven twelfths, or twelve thirteenths, of the world's population of Semper Augustus tulips, that those tulips in some sense owned him—that he'd devoted the better part of his life to advancing their numbers and happiness. But the tulipomania he unwittingly helped fire was an inestimable boon to the genus *Tulipa*, which may be said to have had the last laugh. Its fortunes, at least, have been ascendant in the world ever since the Dutch burghers lost their fortunes on its account.

Witting or not, all these characters have been actors in a coevolutionary drama, a dance of human and plant desire that has left neither the plants nor the people taking part in it unchanged. Okay, desire might be too strong a word for whatever it is that drives plants to reinvent themselves so that we might do their bidding, but then, our own designs have often been no more willful than the plants'. We too cast unconscious evolutionary votes every time we reach for the most symmetrical flower or the longest french fry. The survival of the sweetest, the most beautiful, or the most intoxicating proceeds according to a dialectical process, a give-and-take between human desire and the universe of all plant possibility. It takes two, but it doesn't take intention, or consciousness.

I keep coming back to that image of John Chapman floating down the Ohio River, snoozing alongside his mountain of apple seeds—seeds that held sleeping within them the apple's American future, the golden age to come. The barefoot crank knew something about how things stand between us and the plants, something we seem to have lost sight of in the two centuries since. He understood, I think, that our destinies on the river of natural history are twined. And while I personally don't think he was right to judge grafting a "wickedness," his judgment does bespeak an instinctive feeling for the necessity of wildness and the value of multiplicity over monoculture. Though Chapman would probably disagree, genetic engineering is probably no more wicked than grafting, though it too wars against wildness and multiplicity (albeit much more fiercely). It too places its bet—a very large bet—on the Apollonian One as against the Dionysian Many.

The NewLeaf marks an evolutionary turn that may or may not take us somewhere we want to be. Just in case it doesn't, though, we'd be wise to follow Chapman's example, to save and seed all manner of plant genes: the wild, the unpatentable, even the seemingly useless, patently ugly, and just plain strange. Next year in place of the NewLeaf I plan to plant a great many different Old Leafs; instead of one perfect potato, I'll make Chapman's bet on the field. To shrink the sheer diversity of life, as the grafters and monoculturists and genetic engineers would do, is to shrink evolution's possibilities, which is to say, the future open to all of us. "This is the assembly of life that it took a billion years to evolve," the zoologist E. O. Wilson has written, speaking of biodiversity. "It has eaten the storms—folded them into its genes—and created the world that created us. It holds the world steady." To risk this multiplicity is to risk unstringing the world.

Biodiversity is a word that was not in John Chapman's vocabulary, though it's a good one to describe the crazy archive of apple genes he had with him that summer afternoon on the Ohio. His view of our place in nature was eccentric even by the standards of his day. But I'm convinced that there is some usable truth there, if not in his words, then certainly in his deeds. I'm thinking specifically of the way he rigged up his canoe that day, the two hulls side by side, so that the weight of the apple seeds balanced the weight of the man, each helping to keep the other steady on the river. Laughable as an example of naval architecture, perhaps, but seaworthy, surely, as a metaphor. Chapman's craft, his example, invites us to imagine a very different kind of story about Man and Nature, one that shrinks the distance between the two, so that we might again begin to see them for what they are and in spite of everything will always be, which is in this boat together.

SOURCES

Listed below, by chapter, are the principal works referred to in the text, as well as others that supplied me with facts or influenced my thinking.

INTRODUCTION: THE HUMAN BUMBLEBEE

David Attenborough's 1995 public television series *The Private Life of Plants* probably did more than any book to open my eyes to the natural and human world as seen from the plant's point of view. The series' brilliant time-lapse photography immediately makes you realize that our sense of plants as passive objects is a failure of imagination, rooted in the fact that plants occupy what amounts to a different dimension.

On the history of domestication and the relationship between plants and people, I found these books particularly illuminating:

Anderson, Edgar. *Plants, Man and Life* (Berkeley: University of California Press, 1952). A classic on the origins of agriculture.

Balick, Michael J., and Paul Alan Cox. *Plants, People and Culture: The Science of Ethnobotany* (New York: Scientific American Library, 1996).

Bronowski, J. *The Ascent of Man* (Boston: Little, Brown, 1973).

Budiansky, Stephen. *The Covenant of the Wild: Why Animals Chose Domestication* (New York: William Morrow, 1992).

Coppinger, Raymond P., and Charles Kay Smith. "The Domestication of Evolution," *Environmental Conservation,* vol. 10, no. 4, Winter 1983, pp. 283–91. This essay puts domestication into the context of evolution, suggesting that what constitutes "fitness" in nature fundamentally changed during the Neolithic era.

Diamond, Jared. *Guns, Germs, and Steel: The Fates of Human Societies*

(New York: W. W. Norton, 1997). Excellent on the history and botany of domestication, why some species participate and others do not.

Eiseley, Loren. *The Immense Journey* (New York: Vintage Books, 1959). As much myth as science, this book manages to dramatize the rise of the angiosperms.

Nabhan, Gary Paul. *Enduring Seeds: Native American Agriculture and Wild Plant Conservation* (San Francisco: North Point Press, 1989).

On the wider subject of evolution and natural selection:

Darwin, Charles. *The Origin of Species,* edited by J. W. Burrow (London: Penguin Books, 1968).

Dawkins, Richard. *The Selfish Gene* (New York: Oxford University Press, 1976).

Dennett, Daniel C. *Darwin's Dangerous Idea: Evolution and the Meanings of Life* (New York: Simon & Schuster, 1995).

Goodwin, Brian. *How the Leopard Changed Its Spots: The Evolution of Complexity* (New York: Charles Scribner's Sons, 1994).

Jones, Steve. *Darwin's Ghost: The Origin of Species Updated* (New York: Random House, 1999).

Ridley, Matt. *The Red Queen: Sex and the Evolution of Human Nature* (New York: Penguin Books, 1993).

Wilson, E. O. *The Diversity of Life* (New York: W. W. Norton, 1992).

CHAPTER 1: THE APPLE

Though he probably won't approve of the portrait of his hero I brought back with me, William Ellery (Bill) Jones was as generous, knowledgeable, and companionable a guide to Johnny Appleseed country as anyone could hope for. Bill also introduced me to several other people in Ohio and Indiana who helped me to piece together Chapman's elusive story: Steven Fortriede at the Allen County Public Library in Fort Wayne; Myrtle Ake, who showed me the Chapman family graveyard in Dexter City; and David Ferre, a pomologist with the Ohio Agricultural Research and Development Center.

The literary and historical record on John Chapman is remarkably thin. The indispensable source on Chapman's life remains Robert Price's 1954 biography, *Johnny Appleseed: Man and Myth* (Gloucester, Mass.:

Peter Smith, 1967). Also indispensable is the 1871 account of Chapman's life published by *Harper's New Monthly Magazine* (vol. 43, pp. 6–11). For showing me that Chapman was a historical figure worth taking seriously, I owe a debt to Edward Hoagland's excellent *American Heritage* profile, "Mushpan Man," which is reprinted in Hoagland's essay collection *Heart's Desire* (New York: Summit Books, 1988). For contemporary accounts of Chapman, I highly recommend *Johnny Appleseed: A Voice in the Wilderness,* an anthology of historical writings on Chapman edited by William Ellery Jones (West Chester, Pa.: Chrysalis Books, 2000). Also worth reading are Chapman's obituary in the Fort Wayne *Sentinel* (March 22, 1845) and Steven Fortriede, "Johnny Appleseed: The Man Behind the Myth," *Old Fort News* (vol. 41, no. 3, 1978).

On the botany, culture, and history of the apple, I profited from interviews and conversations with Bill Vitalis, formerly of the Ellsworth Hill Orchard in Connecticut; Clay Stark and Walter Logan at Stark Brothers Nurseries in Missouri; Tom Vorbeck at Applesource in Illinois; Terry and Judith Maloney at West County Cider in Massachusetts; and, at the USDA Experiment Station in Geneva, New York, Phil Forsline, Herb Aldwinckle, and Susan Brown.

These books on apples, sweetness, and environmental history were particularly helpful:

Beach, S. A. *The Apples of New York* (Albany: J. B. Lyon Company, 1905).

Browning, Frank. *Apples* (New York: North Point Press, 1998). Browning, an orchardist and journalist, traveled to Kazakhstan, visiting the apple's center of diversity with Aimak Djangaliev.

Carlson, R. F., et al. *North American Apples: Varieties, Rootstocks, Outlook* (East Lansing: Michigan State University Press, 1970).

Childers, Norman F. *Modern Fruit Science* (New Brunswick, N.J.: Rutgers University Press, 1975).

Crosby, Alfred. *Ecological Imperialism: The Biological Expansion of Europe, 900–1900* (Cambridge, England: Cambridge University Press, 1986). The preeminent environmental historian on the exchange of species between the Old World and New after Columbus.

———. *Germs, Seeds & Animals: Studies in Ecological History* (Armonk, N.Y.: M. E. Sharpe, 1994).

Haughton, Claire Shaver. *Green Immigrants: The Plants That Transformed America* (New York: Harcourt Brace Jovanovich, 1978).

Marranca, Bonnie, ed. *American Garden Writing* (New York: PAJ Publications, 1988).

Martin, Alice A. *All About Apples* (Boston: Houghton Mifflin, 1976).

Mintz, Sidney W. *Sweetness and Power* (New York: Penguin Books, 1986).

Terry, Dickson. "The Stark Story: Stark Nurseries 150th Anniversary," special issue of the *Bulletin of the Missouri Historical Society,* September 1966.

Thoreau, Henry David. "Wild Apples," in *The Natural History Essays,* introduction and notes by Robert Sattelmeyer (Salt Lake City: Peregrine Smith Books, 1980).

Weber, Bruce. *The Apple in America: The Apple in 19th Century American Art* (New York: Berry-Hill Galleries, 1993). An exhibition catalog.

Yepson, Roger. *Apples* (New York: W. W. Norton, 1994).

On the subjects of Dionysus and Apollo (which also figure in subsequent chapters), I've relied primarily on Friedrich Nietzsche's *The Birth of Tragedy* (London: Penguin Books, 1993; first published 1872) and Camille Paglia's *Sexual Personae* (New Haven: Yale University Press, 1990), a book brimming with insight for anyone who would write or think about nature. The following books were also helpful on Dionysus:

Dodds, E. R. *The Greeks and the Irrational* (Berkeley: University of California Press, 1951).

Frazer, Sir James. *The New Golden Bough* (New York: New American Library, 1959).

Harrison, Jane. *Prolegomena to the Study of Greek Religion* (Cambridge, Mass.: Harvard University Press, 1922).

Kerenyi, Carl. *Dionysus: Archetypal Image of Indestructible Life,* trans. by Ralph Manheim (Princeton: Princeton University Press, 1976).

Otto, Walter F. *Dionysus: Myth and Cult,* trans. by Robert B. Palmer (Bloomington: Indiana University Press, 1965).

Williams, C. K., trans. *The Bacchae of Euripides* (New York: Farrar, Straus and Giroux, 1990).

CHAPTER 2: THE TULIP

On flowers in general, I consulted:

Goody, Jack. *The Culture of Flowers* (Cambridge, England: Cambridge University Press, 1993).

Huxley, Anthony. *Plant and Planet* (London: Penguin Books, 1987).

Proctor, Michael, et al. *The Natural History of Pollination* (Portland, Ore.: Timber Press, 1996).

On the biology and philosophy of beauty:

Etcoff, Nancy. *Survival of the Prettiest* (New York: Doubleday, 1999).

Nietzsche, Friedrich. *The Birth of Tragedy,* op. cit.

Paglia, Camille. *Sexual Personae,* op. cit.

Pinker, Steven. *How the Mind Works* (New York: W. W. Norton, 1997).

Ridley, Matt. *The Red Queen,* op. cit.

Scarry, Elaine. *On Beauty and Being Just* (Princeton: Princeton University Press, 1999).

Turner, Frederick. *Beauty: The Value of Values* (Charlottesville: University Press of Virginia, 1991).

―――. *Rebirth of Value: Meditations on Beauty, Ecology, Religion, and Education* (Albany: State University of New York Press, 1991).

On tulips and the Dutch tulipomania, my principal source was Anna Pavord's definitive and beautiful book, *The Tulip: The Story of a Flower That Has Made Men Mad* (London: Bloomsbury, 1999). Also helpful were:

Baker, Christopher, and Willem Lemmers, Emma Sweeney, and Michael Pollan. *Tulipa: A Photographer's Botanical* (New York: Artisan, 1999).

Chancellor, Edward. *Devil Take the Hindmost: A History of Financial Speculation* (New York: Farrar, Straus and Giroux, 1999). Chancellor is especially good tracing the parallels between market manias and carnivals.

Dash, Mike. *Tulipomania: The Story of the World's Most Coveted Flower and the Extraordinary Passions It Aroused* (New York: Crown, 1999).

Dumas, Alexandre. *The Black Tulip* (New York: A. L. Burt Company, n.d.; first published 1850).

Herbert, Zbigniew. "The Bitter Smell of Tulips," in *Still Life with a Bridge: Essays and Apocryphas* (London: Jonathan Cape, 1993).

Schama, Simon: *The Embarrassment of Riches: An Interpretation of Dutch Culture in the Golden Age* (New York: Vintage Books, 1997).

CHAPTER 3: MARIJUANA

This chapter benefited enormously from interviews, correspondence, and time spent with a handful of people knowledgeable about the science, culture, and politics of cannabis: Allen St. Pierre at NORML; Peter Gorman and Kyle Kushman at *High Times* magazine; David Lenson at the University of Massachusetts; Bryan R., a breeder and grower living in Amsterdam; Valerie and Mike Corral, who grow and give away medical marijuana in Santa Cruz, California; Lester Grinspoon at the Harvard Medical School; John P. Morgan, a pharmacologist at the City University of New York Medical School; Graham Boyd at the ACLU Drug Policy Litigation Project; Rick Musty and his colleagues at the International Cannabis Research Society; Ethan Nadelman and his colleagues at the Lindesmith Center; Allyn Howlett at the Saint Louis University School of Medicine; and Raphael Mechoulam at the Hebrew University in Jerusalem.

These books and articles proved especially enlightening:

Baum, Dan. *Smoke and Mirrors: The War on Drugs and the Politics of Failure* (Boston: Little, Brown, 1996).

Clarke, Robert Connell. *Hashish!* (Los Angeles: Red Eye Press, 1998).

———. *Marijuana Botany* (Berkeley, Calif.: Ronin Publishing, 1981).

De Quincey, Thomas. *Confessions of an English Opium-Eater* (New York: Dover, 1995; first published 1822).

Escohotado, Antonio. *A Brief History of Drugs,* trans. by Kenneth A. Symington (Rochester, Vt.: Park Street Press, 1999).

Fisher, Philip. *Wonder, the Rainbow and the Aesthetics of Rare Experience* (Cambridge, Mass.: Harvard University Press, 1998).

Ginsberg, Allen. "The Great Marijuana Hoax: First Manifesto to End the Bringdown," in *The Atlantic Monthly,* November 1966, pp. 104, 107–12.

Grinspoon, Lester, M.D. *Marihuana Reconsidered* (Oakland, Calif.: Quick American Archives, 1999; first published 1971). Carl Sagan's anonymous marijuana "trip report," attributed to Mr. X, appears here beginning on p. 109. You can also read it—and Allen Ginsberg's article cited above—at Grinspoon's website: www.marijuana-uses.com.*

Huxley, Aldous. *The Doors of Perception and Heaven and Hell* (New York: Perennial Library, 1990; first published 1953).

Institute of Medicine. *Marijuana and Medicine: Assessing the Science Base* (Washington, D.C.: National Academy Press, 1999). Clear, accessible explanation of how cannabinoids work in the brain.

Lenson, David. *On Drugs* (Minneapolis: University of Minnesota Press, 1995). Little known, but one of the most thoughtful and original books ever written on the drug experience. The quote about the romantic imagination is from "The High Imagination," Lenson's Hess lecture at the University of Virginia, April 29, 1999.

McKenna, Terence. *Food of the Gods: The Search for the Original Tree of Knowledge* (New York: Bantam Books, 1992).

Merlin, Mark David. *Man and Marijuana: Some Aspects of Their Ancient Relationship* (Rutherford, N.J.: Fairleigh Dickinson University Press, 1972).

Musty, Richard E., et al., ed. "International Symposium on Cannabis and the Cannabinoids," *Life Sciences,* vol. 56, nos. 23–24, 1995. See also the ICRS website: www.cannabinoidsociety.org.

Nietzsche, Friedrich. "On the Uses and Disadvantages of History for Life," in *Untimely Meditation,* ed. by Daniel Breazeale (Cambridge, England: Cambridge University Press, 1997).

Pinker, Steven. *How the Mind Works,* op. cit.

Plant, Sadie. *Writing on Drugs* (New York: Farrar, Straus and Giroux, 2000).

Schivelbusch, Wolfgang. *Tastes of Paradise: A Social History of Spices, Stimulants, and Intoxicants,* trans. by David Jacobson (New York: Vintage Books, 1992).

Schultes, Richard E. "Man and Marijuana," *Natural History,* vol. 82, no. 7, 1973, pp. 58–63, 80–82.

Siegel, Ronald K. *Intoxication: Life in Pursuit of Artificial Paradise* (New York: Dutton, 1989).

*All URLs cited were accurate as of November 14, 2000.

Szasz, Thomas. *Ceremonial Chemistry: The Ritual Persecution of Drugs, Addicts, and Pushers* (London: Routledge, 1975).

Wasson, E. Gordon, et al. *Persephone's Quest: Entheogens and the Origins of Religion* (New Haven: Yale University Press, 1986). A reasonable and serious work on what is still a highly speculative field of inquiry.

Weil, Andrew. *The Natural Mind: An Investigation of Drugs and the Higher Consciousness* (New York: Houghton Mifflin, 1986; first published 1972). A quarter century after it first appeared, this remains one of the sanest books on drugs.

Zimmer, Lynn, and John P. Morgan. *Marijuana Myths, Marijuana Facts: A Review of the Scientific Evidence* (New York: The Lindesmith Center, 1997).

CHAPTER 4: THE POTATO

This chapter had its origins in an article on Monsanto and genetically modified food I wrote for *The New York Times Magazine* ("Playing God in the Garden," October 25, 1998, pp. 44–50, 51, 62–63, 82, 92–93). While I was researching that article, Monsanto was remarkably open and generous, giving me access to its scientists, executives, laboratories, customers—and seed potatoes. My education in the science and politics of genetic engineering also owes a great deal to Margaret Mellon at the Union of Concerned Scientists; Andrew Kimbrell at the Center for Technology Assessment; Rebecca Goldberg at the Environmental Defense Fund; Betsy Lydon at Mothers & Others; Hope Shand and her colleagues at RAFI; and Steve Talbott's excellent website on technology and society, www.netfuture.org. I also received an invaluable education from the farmers who took the time to talk to me and show me around: Mike Heath, Nathan Jones, Woody Deryckx, Danny Forsyth, Steve Young, and Fred Kirschenmann.

On the botany and social history of the potato, as well as on agriculture generally, I found these books especially helpful:

Anderson, Edgar. *Plants, Man and Life*, op. cit.

Berry, Wendell. *The Gift of Good Land* (San Francisco: North Point Press, 1981). Still the wisest voice on the connections between agriculture and everything else.

———. *Life Is a Miracle: An Essay Against Modern Superstition* (Washington, D.C.: Counterpoint, 2000).

————. *The Unsettling of America: Culture & Agriculture* (San Francisco: Sierra Club Books, 1977).

Diamond, Jared. *Guns, Germs, and Steel,* op. cit.

Fowler, Cary, and Pat Mooney. *Shattering: Food, Politics, and the Loss of Genetic Diversity* (Tucson: University of Arizona Press, 1996).

Gallagher, Catherine, and Stephen Greenblatt. *Practicing New Historicism* (Chicago: University of Chicago Press, 2000). See especially Chapter 4, "The Potato in the Materialist Imagination," written by Gallagher.

Harland, Jack R. *Crops and Man* (Madison, Wis.: American Society of Agronomy, 1992).

Hobhouse, Henry. *Seeds of Changes: Five Plants That Changed Mankind* (London: Harper & Row, 1986).

Holden, John, et al. *Genes, Crops, and the Environment* (Cambridge, England: Cambridge University Press, 1993).

Howard, Sir Albert. *An Agricultural Testament* (London: Oxford University Press, 1940).

Lewontin, Richard. *Biology as Ideology: The Doctrine of DNA* (New York: Harper Perennial, 1991). A skeptical voice on genetic determinism, the orthodoxy of our time.

————. *The Triple Helix: Gene, Organism, and Environment* (Cambridge, Mass.: Harvard University Press, 2000).

Salaman, Redcliffe. *The History and Social Influence of the Potato* (Cambridge, England: Cambridge University Press, 1985; first published 1949). Everything you wanted to know, and then some.

Scott, James C. *Seeing Like a State: How Certain Schemes to Improve the Human Condition Have Failed* (New Haven, Conn.: Yale University Press, 1998). This fascinating multidisciplinary study of government, architecture, and agriculture is indispensable on the subject of monoculture, which Scott puts into the context of modernism.

Shiva, Vandana. *Biopiracy: The Plunder of Nature and Knowledge* (Boston: South End Press, 1997).

————. *Stolen Harvest: The Hijacking of the Global Food Supply* (Boston: South End Press, 2000).

Tilman, David. "The Greening of the Green Revolution," *Nature,* November 19, 1998, pp. 211–12.

Van der Ploeg, Jan Douwe. "Potatoes and Knowledge," in *An Anthropo-*

logical Critique of Development, ed. by Mark Hobart (London: Routledge, 1993).

Viola, Herman J., and Carolyn Margolis, eds. *Seeds of Change: Five Hundred Years Since Columbus* (Washington, D.C.: Smithsonian Instutition Press, 1991). See especially the contributions by Alfred Crosby, William F. McNeill, and Sidney W. Mintz.

Weatherford, Jack. *Indian Givers: How the Indians of the Americas Transformed the World* (New York: Crown Publishers, 1988).

Wilson, E. O. *The Diversity of Life,* op. cit.

Zuckerman, Larry. *The Potato: How the Humble Spud Rescued the Western World* (Boston: Faber & Faber, 1998).

INDEX

ABOUT THE TYPE

This book was set in Minion, a 1990 Adobe Originals typeface by Robert Slimbach. Minion is inspired by classical, old-style typefaces of the late Renaissance, a period of elegant, beautiful, and highly readable type designs. Created primarily for text setting, Minion combines the aesthetic and functional qualities that make text type highly readable with the versatility of digital technology.